Genetic Algorithms: Principles and Perspectives
A Guide to GA Theory

OPERATIONS RESEARCH/COMPUTER SCIENCE INTERFACES SERIES

Series Editors

Professor Ramesh Sharda
Oklahoma State University

Prof. Dr. Stefan Voß
Technische Universität Braunschweig

Other published titles in the series:

Greenberg, Harvey J. / *A Computer-Assisted Analysis System for Mathematical Programming Models and Solutions: A User's Guide for ANALYZE*

Greenberg, Harvey J. / *Modeling by Object-Driven Linear Elemental Relations: A Users Guide for MODLER*

Brown, Donald/Scherer, William T. / *Intelligent Scheduling Systems*

Nash, Stephen G./Sofer, Ariela / *The Impact of Emerging Technologies on Computer Science & Operations Research*

Barth, Peter / *Logic-Based 0-1 Constraint Programming*

Jones, Christopher V. / *Visualization and Optimization*

Barr, Richard S./ Helgason, Richard V./ Kennington, Jeffery L. / *Interfaces in Computer Science & Operations Research: Advances in Metaheuristics, Optimization, & Stochastic Modeling Technologies*

Ellacott, Stephen W./ Mason, John C./ Anderson, Iain J. / *Mathematics of Neural Networks: Models, Algorithms & Applications*

Woodruff, David L. / *Advances in Computational & Stochastic Optimization, Logic Programming, and Heuristic Search*

Klein, Robert / *Scheduling of Resource-Constrained Projects*

Bierwirth, Christian / *Adaptive Search and the Management of Logistics Systems*

Laguna, Manuel / González-Velarde, José Luis / *Computing Tools for Modeling, Optimization and Simulation*

Stilman, Boris / *Linguistic Geometry: From Search to Construction*

Sakawa, Masatoshi / *Genetic Algorithms and Fuzzy Multiobjective Optimization*

Ribeiro, Celso C./ Hansen, Pierre / *Essays and Surveys in Metaheuristics*

Holsapple, Clyde/ Jacob, Varghese / Rao, H. R. / *BUSINESS MODELLING: Multidisciplinary Approaches — Economics, Operational and Information Systems Perspectives*

Sleezer, Catherine M./ Wentling, Tim L./ Cude, Roger L. / *HUMAN RESOURCE DEVELOPMENT AND INFORMATION TECHNOLOGY: Making Global Connections*

Voß, Stefan, Woodruff, David / *Optimization Software Class Libraries*

Upadhyaya et al/ *MOBILE COMPUTING: Implementing Pervasive Information and Communications Technologies*

Reeves, Colin & Rowe, Jonathan/ *GENETIC ALGORITHMS—Principles and Perspectives: A Guide to GA Theory*

GENETIC ALGORITHMS: PRINCIPLES AND PERSPECTIVES
A Guide to GA Theory

COLIN R.REEVES
School of Mathematical and Information Sciences
Coventry University

JONATHAN E.ROWE
School of Computer Science
University of Birmingham

Kluwer Academic Publishers
Boston/Dordrecht/London

Distributors for North, Central and South America:
Kluwer Academic Publishers
101 Philip Drive
Assinippi Park
Norwell, Massachusetts 02061 USA
Telephone (781) 871-6600
Fax (781) 871-9045
E-Mail: kluwer@wkap.com

Distributors for all other countries:
Kluwer Academic Publishers Group
Post Office Box 322
3300 AH Dordrecht, THE NETHERLANDS
Telephone 31 786 576 000
Fax 31 786 576 254
E-mail: services@wkap.nl

 Electronic Services <http://www.wkap.nl>

Library of Congress Cataloging-in-Publication Data

Genetic algorithms : principles and perspectives : a guide to GA theory / Colin R. Reeves, Jonathan E. Rowe.
       p. cm. -- (Operations research/ computer science interfaces series ; ORCS 20)
   Includes bibliographical references and index.
   ISBN 1-4020-7240-6
   1. Genetic algorithms. I. Reeves, Colin R. II. Rowe, Jonathan E. III. Series.

QA402.5 .G456 2002
519.3--dc21

2002032067

Copyright © 2003 by Kluwer Academic Publishers. Second printing 2004.

All rights reserved. No part of this work may be reproduced, stored in a retrieval system, or transmitted in any form or by any means, electronic, mechanical, photocopying, microfilming, recording, or otherwise, without the written permission from the Publisher, with the exception of any material supplied specifically for the purpose of being entered and executed on a computer system, for exclusive use by the purchaser of the work.

Permission for books published in Europe: permissions@wkap.nl
Permissions for books published in the United States of America: permissions@wkap.com

Printed on acid-free paper.

Printed in United Kingdom by Biddles/IBT Global

Contents

1 Introduction — **1**
  1.1 Historical Background — 1
  1.2 Optimization — 4
  1.3 Why GAs? Why Theory? — 9
  1.4 Bibliographic Notes — 17

2 Basic Principles — **19**
  2.1 GA Concepts — 19
  2.2 Representation — 19
  2.3 The Elements of a Genetic Algorithm — 25
  2.4 Illustrative Example — 49
  2.5 Practical Considerations — 51
  2.6 Special Cases — 55
  2.7 Summary — 59
  2.8 Bibliographic Notes — 60

3 Schema Theory — **65**
  3.1 Schemata — 65
  3.2 The Schema Theorem — 68
  3.3 Critiques of Schema-Based Theory — 74
  3.4 Surveying the Wreckage — 78
  3.5 Exact Models — 84
  3.6 Walsh Transforms and deception — 86
  3.7 Conclusion — 90
  3.8 Bibliographic Notes — 91

4 No Free Lunch for GAs — **95**
  4.1 Introduction — 95
  4.2 The Theorem — 96

	4.3	Developments	103
	4.4	Revisiting Algorithms	106
	4.5	Conclusions	108
	4.6	Bibliographic Notes	109
5	**GAs as Markov Processes**		**111**
	5.1	Introduction	111
	5.2	Limiting Distribution of the Simple GA	118
	5.3	Elitism and Convergence	122
	5.4	Annealing the Mutation Rate	127
	5.5	Calculating with Markov Chains	130
	5.6	Bibliographic Notes	138
6	**The Dynamical Systems Model**		**141**
	6.1	Population Dynamics	141
	6.2	Selection	145
	6.3	Mutation	151
	6.4	Crossover	158
	6.5	Representational Symmetry	163
	6.6	Bibliographic Notes	169
7	**Statistical Mechanics Approximations**		**173**
	7.1	Approximating GA Dynamics	173
	7.2	Generating Functions	176
	7.3	Selection	179
	7.4	Mutation	183
	7.5	Finite Population Effects	190
	7.6	Crossover	192
	7.7	Bibliographic Notes	197
8	**Predicting GA Performance**		**201**
	8.1	Introduction	201
	8.2	Epistasis Variance	202
	8.3	Davidor's Methodology	203
	8.4	An Experimental Design Approach	204
	8.5	Walsh representation	207
	8.6	Other related measures	215
	8.7	Reference classes	221
	8.8	General Discussion	223
	8.9	Bibliographic Notes	228

CONTENTS

9 Landscapes — **231**
- 9.1 Introduction 231
- 9.2 Mathematical Characterization 237
- 9.3 *Onemax* Landscapes 246
- 9.4 Long Path Landscapes 253
- 9.5 Local Optima and Schemata 256
- 9.6 Path Tracing GAs 259
- 9.7 Bibliographic Notes 261

10 Summary — **265**
- 10.1 The Demise of the Schema Theorem 265
- 10.2 Exact Models and Approximations 268
- 10.3 Landscapes, Operators and Free Lunches 273
- 10.4 The Impact of Theory on Practice 274
- 10.5 Future research directions 278
- 10.6 The Design of Efficient Algorithms 285

A Test Problems — **287**
- A.1 Unitation 287
- A.2 *Onemax* 288
- A.3 Trap functions 288
- A.4 Royal Road 288
- A.5 Hierarchical-IFF (HIFF) 290
- A.6 Deceptive functions 290
- A.7 Long paths 291
- A.8 NK landscape functions 292
- A.9 ℓ, θ functions 292

Bibliography — **295**

Index — **327**

Preface

As the 21st century begins, genetic algorithms are becoming increasingly popular tools for solving hard optimization problems. Yet while the number of applications has grown rapidly, the development of GA theory has been very much slower. The major theoretical work on GAs is undoubtedly Michael Vose's *The Simple Genetic Algorithm*. It is a magnificent *tour de force*, but it is hardly easy for novices to come to grips with, and it focuses almost exclusively (although with good reason) on the dynamical systems model of genetic algorithms.

However, there have been several other theoretical contributions to genetic algorithms, as evidenced by the papers in the well-established *Foundations of Genetic Algorithms* series of biennial workshops. There really is no book-length treatment of the subject of GA theory, so we have written this book in the conviction that the time is ripe for an attempt to survey and synthesize existing theoretical work, with the intent of preparing the way for further theoretical advances. Actually, a synthesis is not here yet—the field is still a little too fragmented, which is why we have described this book as a set of perspectives. But there are encouraging signs that the different perspectives are beginning to integrate, and we hope that this book will promote the process.

It has been our endeavour to make this material as accessible as possible, but inevitably some mathematical knowledge is necessary to gain a full understanding of the topics covered. Perhaps the most important sets of mathematical concepts needed fully to appreciate GA theory are those of linear algebra and stochastic processes, and students with such a background will be better equipped than most to make a contribution to the subject. Nevertheless, we have tried to explain necessary concepts in such a way that much of this book can be understood at advanced undergraduate level, although we expect it to be of most relevance to post-graduates.

Acknowledgements

Clearly, many others have contributed in no small way to this book. We would like to thank our colleagues at Coventry and Birmingham for their interest and support. Some of the material has been used as part of a Masters programme at Coventry, and several students have made helpful comments. We also owe a debt to many friends and colleagues in the evolutionary algorithms community with whom we have discussed and debated some of these topics over the years: in particular, Ken De Jong, Bart Naudts, Riccardo Poli, Adam Prügel-Bennett, Günter Rudolph, Michael Vose, Darrell Whitley and Alden Wright.

Colin Reeves and Jon Rowe
July 2002

About the Authors

Colin Reeves obtained his BSc in Mathematics followed by research for the degree of MPhil at what is now Coventry University. After working for the Ministry of Defence as an OR scientist, he returned to Coventry to lecture in Statistics and OR. There he was awarded a PhD in recognition of his published works, and later became Professor of Operational Research. His main research interests are in pattern recognition, in heuristic methods for combinatorial optimization, most particularly in evolutionary algorithms. His book *Modern Heuristic Techniques for Combinatorial Optimization* has been widely used, and he has published many papers on evolutionary algorithms, with a strong bias towards foundational issues. He co-chaired the 1998 *Foundations of Genetic Algorithms* workshop, and is an Associate Editor of both *Evolutionary Computation* and *IEEE Transactions on Evolutionary Computation*.

Jon Rowe received a BSc in Mathematics and a PhD in Computer Science from the University of Exeter. He held research positions in industry, at Buckingham University and then De Montfort University. He is currently a Lecturer in Computer Science at the University of Birmingham. His main research interests are in the theory of genetic algorithms, and emergent phenomena in multi-agent systems. He is co-chair of the 2002 *Foundations of Genetic Algorithms* workshop.

Chapter 1

Introduction

In the last 20 years, the growth in interest in heuristic search methods for optimization has been quite dramatic. Having once been something of a Cinderella in the field of optimization, heuristics have now attained considerable respect and are extremely popular. One of the most interesting developments is in the application of genetic algorithms (GAs). While other techniques, such as simulated annealing (SA) or tabu search (TS), have also become widely known, none has had the same impact as the GA, which has regularly been featured, not merely in the popular scientific press, but also in general newspaper articles and TV programs.

There are now several excellent books available that will introduce the reader to the concepts of genetic algorithms, but most of these are content to use fairly conventional accounts of what GAs are, and how they work. These rest on the pioneering work of John Holland [124] and his development of the idea of a schema. However, in the past 10 years some significant research has been carried out that has cast some doubt on the universal validity of schema theory, and several different theoretical perspectives have been developed in order to help us understand GAs better. These ideas are somewhat fragmented, and have not previously been collected in book form. This is the purpose of this work.

1.1 Historical Background

The term *genetic algorithm*, almost universally abbreviated nowadays to GA, was first used by Holland [124], whose book *Adaptation in Natural and Artificial Systems* of 1975 was instrumental in creating what is now a flourishing field of research and application that goes much wider than the original GA. The subject now includes evolution strategies (ES), evolutionary programming (EP), artificial life (AL), classifier systems (CS), genetic programming (GP), and most recently the concept of evolvable hardware. All these related fields of research are often nowadays grouped under the heading of *evolutionary computing* or *evolutionary algorithms* (EAs). While this book concentrates on the GA, we would not wish to give the impression that these other topics are of no importance: it is just that their differences tend to give rise to different problems, and GA theory in the restricted sense is hard enough. For example, ES tends to focus on continuous variables, which requires a different theoretical approach; GP nearly always involves variable length representations while GAs nearly always assume a fixed length.

While Holland's influence in the development of the topic has been incalculable, from an historical perspective it is clear that several other scientists with different backgrounds were also involved in proposing and implementing similar ideas. In the 1960s Ingo Rechenberg [211] and Hans-Paul Schwefel [260] in Germany developed the idea of the *Evolutionsstrategie* (in English, *evolution strategy*), while—also in the 1960s—Lawrence Fogel [81, 82] and others in the USA implemented their idea for what they called *evolutionary programming*. What these proposals had in common were their allegiance to the ideas of mutation and selection that lie at the centre of the neo-Darwinian theory of evolution. Only Bremermann [29] and Fraser [86] considered the use of recombination—the idea later placed at the heart of GAs by John Holland. However, although some promising results were obtained, as an area of research the topic of evolutionary computing did not really catch fire. There may be many reasons for this, but one that is surely relevant is that the techniques tended to be computationally expensive—unless the applications were to be trivial ones—and in the 1960s the power of computers was many orders of magnitude less than it is today. Nevertheless, the work of these early explorers is fascinating to read in the light of our current knowledge; David Fogel (son of one of these pioneers, and now with a considerable reputation of his own) has done a great service to the EA community in tracking down and collecting some of this work in his aptly titled book *Evolutionary Computing: the Fossil Record* [79].

In any mention of evolution, we have to reckon with Charles Darwin [44]. His *Origin of Species* is certainly one of the most famous books in the world. His theory introduced the idea of *natural selection* as the mechanism whereby small heritable variations in organisms could accumulate under the assumption that they induce an increase in fitness—the ability of that organism to reproduce. What might cause such variations was something about which Darwin could only speculate. Shortly after the publication of the

1.1. HISTORICAL BACKGROUND

Origin of Species in 1859, but unknown to almost everyone in the scientific world for another 30 years, Gregor Mendel discovered the genetic basis of inheritance. Later, scientists such as Hugo de Vries developed the concept of genetic mutations, and in the first half of the 20th century, the newer discoveries were integrated into Darwin's theory in what became known as the *modern synthetic theory of evolution*, often called *neo-Darwinism*. This synthesis emerged after a long debate over the importance of selection relative to other influences upon natural populations such as genetic drift and geographical isolation. Provine [194] gives an absorbing account of these debates through the medium of a biography of Sewall Wright, one of the protagonists who was also at the forefront of developments in theoretical population genetics. According to neo-Darwinism, the cumulative selection of 'fitter' genes in populations of organisms leads over time to the separation into new species, and over really long periods of time, this process is believed to have led to the development of higher taxa above the species level. Probably the best-known current defender of 'pure' neo-Darwinism is Richard Dawkins [52].

An oft-quoted example of neo-Darwinism in action is the peppered moth *Biston betularia*. Before the industrial revolution in the UK the common form of this moth (known as *typica*) was light in colour with small dark spots—good camouflage, on the light background of a lichen-covered tree trunk, against predators. With the increase of pollution the tree trunks became gradually darker, and a mutated dark form *carbonaria* became more prevalent. Natural selection in the form of bird predation was presumed to account for the relative loss of fitness of the common form, and the consequent spread of *carbonaria*. Recently, the *typica* form has become more widespread again—presumably as a by-product of pollution control legislation.

In imitation of this type of process, a GA can allow populations of potential solutions to optimization problems to die or reproduce with variations, gradually becoming adapted to their 'environment'—in this case, by means of an externally imposed measure of 'fitness'.

1.1.1 Recombination

Holland followed biological terminology fairly closely, describing the encoded individuals that undergo reproduction in his scheme as *chromosomes*, the basic units of those individuals as *genes*, and the set of values that the genes can assume as *alleles*. The position of a gene in a string is called its *locus*—in GA practice, a term that is often used interchangeably with the term gene.

The distinction between the *genotype*—the actual genetic structure that represents an individual, and the *phenotype*—its observed characteristics as an organism, has also become a popular analogy.

While the other pioneers in evolutionary computation had mainly emphasized mutation and selection in accordance with neo-Darwinian orthodoxy, Holland introduced a distinctive focus on *recombination*, the importance of which to evolutionary theory has grown with the unravelling of the structure of DNA and the increased understanding of genetics at the molecular level. For example, the view of the eminent evolutionary biologist Ernst Mayr is that

> *genetic recombination is the primary source of genetic variation.*
> [159]

In Holland's genetic algorithm, not only do the individuals mutate into new ones, but there is also an alternative route to the generation of new chromosomes whereby pairs of individuals can exchange some of their genes. This is accomplished by a process called *crossover*.

Despite this promotion of the value of recombination as well as mutation, it cannot be said that the value of Holland's work initially gained any speedier recognition than that of the other EC pioneers, even in the 1970s when his book was first published. However, in 1975 another important work appeared: Ken De Jong, a graduate student of Holland's, completed his doctoral thesis [55]. What was distinctive about this was in its focus on optimization problems, particularly his careful selection of a varied set of test problems.[1] Here at last was an emphasis that spoke to an existing research community eager for new tools and techniques.

1.2 Optimization

In speaking of optimization, we have the following situation in mind: there exists a search space $\mathcal{V}$, and a function

$$g : \mathcal{V} \mapsto \mathbb{R},$$

and the problem is to find

$$\arg\min_{v \in \mathcal{V}} g.$$

---

[1]These test problems have entered GA folklore as the 'De Jong test suite'. For some time, no GA research was complete if it failed to use these benchmarks. While it is now recognized that they do not really cover the spectrum of all possible problems, they remain historically and methodologically important.

1.2. OPTIMIZATION

Here, v is a vector of *decision variables*, and g is the *objective function*. In this case we have assumed that the problem is one of minimization, but everything we say can of course be applied *mutatis mutandis* to a maximization problem. Although specified here in an abstract way, this is nonetheless a problem with a huge number of real-world applications.

In many cases the search space is discrete, so that we have the class of *combinatorial optimization problems* (COPs). When the domain of the function g is continuous, a different approach may well be required, although even here we note that in practice, optimization problems are usually solved using a computer, so that in the final analysis the solutions are represented by strings of binary digits (bits). While it certainly makes the analysis of GAs rather easier if this is assumed (and we shall therefore frequently assume it), we should point out that binary representation is not an essential aspect of GAs, despite the impression that is sometimes given in articles on the subject.[2]

Optimization has a fairly small place in Holland's work on adaptive systems, yet this has now in some ways become the *raison d'être* of GAs, insofar as the majority of research in this area appears to be concentrated on this aspect of their behaviour. One only has to peruse the proceedings of the various conferences devoted to the subject of GAs, or to scan the pages of the journals *Evolutionary Computation* and the *IEEE Transactions on Evolutionary Computing*, as well as most of the mainstream operational research journals, to see that solving optimization problems using GAs has become a very popular sport indeed.

De Jong, who might fairly be credited with initiating the stress on optimization, has cautioned that this emphasis may be misplaced in a paper [58] in which he contends that GAs are not really function optimizers, and that this is in some ways incidental to the main theme of adaptation. Nevertheless, optimization remains a potent concept in GA research, and one gets the impression that it is unlikely to change.

One very strong reason for this is that, in promoting the idea of recombination, GAs appear to have hit upon a particularly useful approach to optimizing complex problems. It is interesting in this regard to compare some of the ideas being put forward in the 1960s in a rather different field—that of *operational research* (OR).

---

[2] Holland's early work stressed binary representations, to the extent that they were believed to possess certain properties that made bit strings in some sense 'optimal'. However, later work has strongly questioned this point, as will be seen in Chapter 3, and with the storage available on modern computers to represent continuous values to a very high degree of precision, there is really no longer a pragmatic argument either.

OR developed in the 1940s and 1950s originally as an interdisciplinary approach for solving complex problems. The impetus was initially provided by the exigencies of war, and subsequently the techniques developed were taken up enthusiastically by large-scale industry—in the UK, especially by the newly-nationalised energy and rail transport industries. The development by George Dantzig of the famous *simplex* algorithm for solving linear programming (LP) problems[3] led to great advances in some areas of industry. However, it was frustratingly clear that many problems were too complex to be formulated in this way—and even if to do so was possible, there was a substantial risk that the 'solution' obtained by linearization might be quite useless in practice.

There was thus a need to develop solution techniques that were able to solve problems with non-linearities, but such methods were frequently too computationally expensive to be implemented, and OR workers began to develop techniques that seemed able to provide 'good' solutions, even if the quality was not provably optimal (or even near-optimal). Such methods became known as *heuristics*. A comprehensive introduction to heuristic methods can be found in [213].

One of the problems that many researchers tried to solve has become recognized as the archetypal combinatorial optimization problem—the travelling salesman problem (usually abbreviated to TSP). This famous problem is as follows: a salesman has to find a route which visits each of n cities once and only once, and which minimizes the total distance travelled. As the starting point is arbitrary, there are clearly $(n-1)!$ possible solutions (or $(n-1)!/2$ if the distance between every pair of cities is the same regardless of the direction of travel). Solution by complete enumeration of the search space (the set of all permutations of the numbers $1, 2, \ldots, n$) is clearly impossible, so large does it become even for fairly small values of n. The *time complexity* of complete enumeration, in other words, grows at an enormous rate. Better techniques are possible, but their time complexity still tends to grow at a rate that is an exponential function of problem size. While methods have been found for some other COPs whose time complexity is merely polynomial, most computer scientists and mathematicians doubt that algorithms ever will be found for a large class of combinatorial optimization problems—those known as NP-hard. This class contains many well-known

---

[3]LP problems are those that can be formulated in such a way that the function g is linear, while the search space $\mathcal{V}$ is a simplex defined by a set of linear constraints. Although the decision variables are assumed to be continuous, the problem turns out to be combinatorial, in that the candidate optimal solutions form a discrete set composed of those points which lie at the corners of the simplex.

1.2. OPTIMIZATION

and well-studied COPs, and it would be a major surprise if polynomial-time algorithms were to be discovered for problems in this class.

The topic of computational complexity *per se* is not a major concern of this book. For a simple introduction, see the discussion in [213]; for an in-depth treatment, probably the best guide is still [88]. However, the importance of complexity research as a motivating factor for the development of GAs for optimization should not be underestimated: the publication of the work of Holland and De Jong dovetailed so well with the need of the OR community for strong generic heuristic methods.

One of the earliest heuristics was the attempt by Croes [40] to solve the TSP. He recognized that no optimal tour would cross itself (at least in Euclidean spaces). Implementing this idea on a computer (which cannot 'see' a tour in the same way that a human can) meant asking it to check all pairs of links to see if they could be re-connected in a way that shortened the tour. When no pair of links can be re-connected without lengthening the tour, it is clear that the tour does not cross itself, and the process terminates. Of course, the tour is not necessarily the best one, but it belongs to a class of relatively 'good' solutions. This idea can readily be generalized to a technique known as *neighbourhood search* (NS), which has been used to attack a vast range of COPs, and we shall have much more to say on the subject later. For the moment, we merely display a generic version of NS in Table 1.1.

In 1965, one of the most influential papers in the history of heuristic search was published by Lin [152], who extended Croes's idea by examining the effect of breaking and re-connecting all groups of 3 links. Croes's solutions he called '2-optimal'; those produced by Lin's own procedure were dubbed '3-optimal'. Empirically, he found that 3-optimal solutions were far superior, and in the case of the (rather small) problems he investigated, often close to the global optimum. However, there is another aspect of that paper to which we wish to draw attention.

Lin also discovered that while starting with different initial permutations gave different 3-optimal solutions, these 3-optimal solutions tended to have a lot of features in common. He therefore suggested that search should be concentrated on those links about which there was not a consensus, leaving the common characteristics of the solutions alone. This was not a GA as Holland was developing it, but there are clear resonances. Much later, after GAs had become more widely known, Lin's ideas were re-discovered as 'multi-parent recombination' and 'consensus operators'.

Other OR research of the same era took up these ideas. Roberts and Flores [237] (apparently independently) used a similar approach to Lin's for

---

Generate an initial solution v;
Specify a neighbourhood function $N(v)$;
Store v as current best v^o and evaluate $g^o = g(v^o)$;
while termination condition not satisfied **do**
    select a solution $v' \in N(v)$;
    evaluate $g' = g(v')$;
    if $g' < g^o$ then
        store v' as current best v^o and g' as g^o;
    set $v = v'$;
endwhile
Output v^o and g^o.

Table 1.1: A neighbourhood search template for minimization. The type of neighbourhood, and the specific criteria for selecting neighbours and for ending the search, will determine the search trajectory in any particular case. By specifying these criteria in different ways, the method will yield a variety of procedures. Perhaps the most common case is that of *steepest descent*, in which all neighbours are evaluated and the one that most improves the current $g(v)$ value is selected; it terminates when no improving solutions can be found.

the TSP, while Nugent *et al.* [183] applied the same ideas to the quadratic assignment problem. However, the general principle was not adopted into OR methodology, and relatively little was done to exploit the idea until GAs came on the OR scene in the 1990s. Nevertheless, this represents an interesting development that is little known in GA circles, and will provide an interesting route into one of the perspectives on GAs that we shall discuss later in this book.

1.3 Why GAs? Why Theory?

In this book, we shall present quite a lot of theory relating to GAs, and we shall try to do so in a relatively accessible manner. However, there are two related questions that the reader might well wish to ask: why use GAs, and why do we need theory?

There are other heuristic approaches that have met with considerable success in applications to optimization problems. Some of these are described in [1, 213, 220]. In the case of another popular technique—that of *simulated annealing* (SA)—there is a considerable body of theoretical analysis that has established how SA works, which is also useful in giving guidelines for its implementation [283]. In the case of GAs, we have as yet nothing comparable that could help us decide when to use a GA, and in what way to implement one, yet the popularity of GAs seems to outweigh that of other techniques such as SA. While these other methods are not unknown to a wider public, GAs appear much more prominent. They have been regularly featured in newspaper articles and TV programs (at least in Europe). It is not possible to give exact figures, but it would appear that the level of funding for GA-based projects by national governments, and by supranational government organizations such as the European Commission, far exceeds any funding given to projects based on methods such as SA or TS.

This disparity must surely be at least partly due to the seductive power of neo-Darwinism. Theodosius Dobzhansky, one of the greatest experimental evolutionary scientists of the 20th century, put it this way:

> *After [Darwin] the fateful idea [of evolution] has become one of the cornerstones on which the thinking of civilised man is based...* [62]

and he later quotes approvingly the words of the Jesuit thinker Teilhard de Chardin

> *[Evolution] is a general postulate to which all theories, all hypotheses, all systems must henceforth bow and which they must satisfy in order to be thinkable and true. Evolution is a light which illuminates all facts, a trajectory which all lines of thought must follow...*[62]

This leads to our second question: why theory? The temptation is to forget about theory altogether. It is not uncommon in popular presentations of the idea of a GA to meet this attitude, where the use of GAs is justified by what is essentially a hand-waving argument: GAs are equated to Darwinian evolution, and there is therefore a sort of inevitability about their potential for all kinds of tasks. Because we're here, the argument seems to be, 'evolution' clearly works, and so imitating it (however approximately) must be a sensible thing to do. For example, one recent book states the following:

> *The GA works on the Darwinian principle of natural selection.... Whether the specifications be nonlinear, constrained, discrete, multimodal, or even NP-hard, GA is entirely equal to the challenge.*

This is a very strong claim to make, considering the long history of optimization, especially in its insouciant reference to NP-hard problems, and ought not to be made without some justification. However, although the authors go on to discuss (rather perfunctorily) some of Holland's ideas about GAs, one feels that this is somewhat superfluous: to invoke the name of Darwin is really all the justification that is needed! Another book puts it this way:

> *The concept that evolution ... generated the bio-diversity we see around us today is a powerful ... paradigm for solving any complex problem.*

These are both books for scientists and engineers which do make an attempt to discuss briefly some ideas as to how GAs work.[4] However, in neither case is there any real rigour in the attempt. Even John Holland himself has occasionally made unguarded statements that tend to encourage grand claims and hand-waving explanations:

> *[A genetic algorithm] is all but immune to some of the difficulties— false peaks, discontinuities, high dimensionality, etc.—that commonly attend complex problems.* [125]

---

[4]The identities of these books have deliberately not been revealed. There is no intention here to attack individual authors. Rather we are making the general point that even among scientists who have more than a nodding acquaintance with GAs, there seems to be a lack of curiosity about the details.

1.3. WHY GAS? WHY THEORY?

This type of approach raises several issues. First of all, one could make the point that copying a process that has apparently taken billions of years does not seem to be a sensibly efficient way of solving complex optimization problems in an acceptable time frame! (This objection was apparently made to Holland early in the development of GAs.) More seriously, evolutionary *mechanisms* are nowhere nearly so well understood as would seem to be the case from reading popular accounts of evolution. The concept itself seems infinitely adaptable, so that almost any account can be given of why a particular mutation could have a selective advantage, and almost any mechanism could be proposed as that which accomplishes 'natural' selection. Yet many of these accounts are highly speculative *post hoc* arguments, not far removed categorically from the 'Just-so Stories' associated with Rudyard Kipling. Gould and Lewontin [104] have in fact made precisely this criticism of what passes for science in much writing on evolution. Although many adaptationist stories are perfectly plausible, it is one thing to conjecture a possible scenario, quite another to demonstrate that it has actually happened.[5]

There are some places where the ground appears firmer, as in that famous textbook example, the peppered moth, as described above. However, it now appears that even this is far less well understood than the textbooks would have us to believe. Recent work by Majerus [156] and by Sargent *et al.* [252] has cast doubt on the original experiments by Kettlewell [146], which had apparently established that natural selection by bird predation was the key. As a review of Majerus's book in *Nature* puts it:

> ..[the peppered moth story] shows the footprints of natural selection, but we have not yet seen the feet. [39].

Moreover, according to [252], the historical aspects of the story are not certain either. The interpretation of the dark form *carbonaria* as a mutation seems to be a conjecture based on the fact that few had been observed before the 1850s. There is no direct evidence that this form did not already exist before the rise of industry. Thus, what has been observed may not really be a speciation event, but merely a fluctuation in the frequencies of two existing species brought about by some as yet unknown mechanism in response to some environmental changes. In a sense it is not surprising that caution is needed: we are talking about explanations for historical

---

[5]Dawkins is scathing about certain scientists adopting what he calls the *argument from personal incredulity*—'I don't see how that could happen, therefore it didn't'. But this invites the obvious retort that he suffers from the opposite affliction, the *argument from personal credulity*—'I can see how it could happen, therefore it did'.

events, and proposing solutions that cannot be completely tested—not that this stops the proliferation of adaptationist stories in the popular scientific literature.

It is also true that while neo-Darwinism, with its almost exclusive emphasis on natural selection, remains the most popular overall narrative, the objections raised in earlier generations [194] have refused to lie down quietly. Reid [233] discusses some of the problems that have not gone away (relating to questions of the scope, rate and possible limitations on the efficacy of natural selection) as he saw it in the mid-1980s, and since then there has been a steady stream of doubters. Starting in the 1950s, there have been enormous strides in molecular biology and developmental biology in particular, and biologists from those backgrounds are certainly less than completely convinced of the all-embracing role of natural selection. As we have discovered more about the genome, the mind-boggling complexity of living systems has become all too apparent, and 'beanbag genetics' (the phrase is Ernst Mayr's) fails to provide a satisfactory story.

While the general outline of a sequence of adaptive changes can be imagined for almost any biological entity, attempts to understand in details the steps required for the development of a complex system at the molecular level are very thin on the ground. Although Dawkins asserts flatly of organisms that

> *it isn't true that each part is essential for the success of the whole*[52] ,

others with recent personal experience of molecular biological research have begged to differ. In a very provocative but stimulating book [18], the molecular biochemist Michael Behe has raised the question of how what he calls *irreducibly complex* molecular biological systems could have come into being at all by the sole means of mutation and natural selection. By 'irreducibly complex', he means systems or processes where the whole is more than the sum of the parts, so that, *contra* Dawkins, a part of the system would be of no selective value to the organism on its own. He gives several interesting examples of such systems, and found nothing in any of the professional literature that comes close to a rigorous explanation for their evolutionary origin.[6] James Shapiro writes similarly as to the questions raised by molecular biology:

---

[6]Predictably, his work has been not been kindly received by the neo-Darwinian 'establishment', and a number of mechanisms have been suggested for his test cases. But so far these are merely the 'argument from personal credulity' again: it does not appear that they do much more than demonstrate a good imagination on behalf of the protagonists.

1.3. WHY GAS? WHY THEORY?

> *The molecular revolution has revealed an unanticipated realm of complexity and interaction ... Localized random mutation, selection operating 'one gene at a time'...and gradual modification of individual functions are unable to provide satisfactory explanations for the molecular data... The point of this discussion is that our current knowledge of genetic change is fundamentally at variance with neo-Darwinist postulates...* [261]

Shapiro is well aware of what this means for our understanding of evolutionary mechanisms, for he continues:

> *Inevitably, such a profound advance in awareness of genetic capabilities will dramatically alter our understanding of the evolutionary process. Nonetheless, neo-Darwinist writers like Dawkins continue to ignore or trivialize the new knowledge and insist on gradualism as the only path for evolutionary change.* [261]

The place of natural selection in the evolutionary scheme has also been questioned by researchers involved in developmental aspects of biology. Lila Gatlin has argued that it is too often invoked to disguise our ignorance:

> *The words 'natural selection' play a role in the vocabulary of the evolutionary biologist similar to the word 'God' in ordinary language.*[89]

Certainly, in many public statements by scientific celebrities, phrases such as the 'imperatives of Darwinian natural selection'[7] trip neatly off the tongue in the same way that in earlier centuries the great and the good would speak of 'divine Providence'. The developmental biologist Brian Goodwin is another of those who is rather impatient with this approach, as the provocative title of his article [102] reveals, and he suggests that other mechanisms (not as yet fully understood) must be at work. The pathway from the genotype to the living organism (phenotype) is incredibly complex, and his book [103] explains at some length what the consequent problems are, and proposes some tentative solutions. Stuart Kauffman [144, 145] also doubts that natural selection is adequate by itself, and his investigations of complexity lead to similar alternative proposals to those of Goodwin, while James Shapiro,

---

Behe's fundamental point that no sound experimental and theoretical work has yet been done to provide an explanation for 'irreducible complexity' seems to remain intact. This much is conceded in a temperate and thoughtful review by Weber [299].

[7]This particular phrase was a favourite with a recent BBC Reith Lecturer.

whose comments were quoted above, has discussed the effect of the developments in molecular biology in similar terms [262].

Thus, if natural selection (in alliance with recombination and mutation) is not able to bear the weight of explaining the biological complexities of the world, it is surely too glib to appeal to it as a justification for using 'evolutionary' techniques for optimization as well. In fact a recent review by Peter Ross argues that a dispassionate view of the evidence suggests the reverse:

> *More than a century of research on evolution has seen the gradual dismantling of the idea that competition and selection drives a continuing improvement...it is not clear that biological evolution is optimizing any kind of global measure of anything.* [241]

Ross's argument, in the passage from which this quotation was taken, is one of critical importance to optimization: that evolution has not been shown actually to be optimizing a 'fitness function' or searching a 'fitness landscape', although this is a common metaphor—it was first advanced in a well-known paper by Sewall Wright [318], and continues to be popular, as the title of another of Dawkins's books [53] suggests. However, in population genetics, *fitness* is usually just a measure of the change in gene frequency in a population—i.e., a measure of *a posteriori* reproductive success.[8] However, in the practical implementation of a GA, we cannot proceed without some *a priori* external measure of fitness of an individual, and it is not clear exactly where the analogy lies in the natural world. This, it appears, is really the problem in the peppered moth debate: clearly one type of moth survives and reproduces better than the other, but exactly *what* constitutes its propensity to do so is still not clear. Even if we do grant the concept of a fitness landscape as a useful metaphor, we also have to realize that the survival and reproduction of real organisms depends on many factors in their environment, and ones that are dynamically changing as well. Further, to some extent, the environment may be manipulated by the actions of the organisms themselves and others. Even the simplest ecosystem is a hugely complicated affair.

Dawkins [51] goes into the question of fitness in some detail, discussing 5 different definitions of fitness in biology. Most of these have to do with some aspect of reproductive success (measured *a posteriori*); only in his first sense

---

[8]The *Oxford Dictionary of Biology*, for example, defines fitness as 'The condition of an organism that is well adapted to its environment, *as measured by its ability to reproduce itself*'. [our italics]

1.3. WHY GAS? WHY THEORY?

is there some resemblance to what GA users mean by fitness. Elsewhere [52] he finesses the problem by smuggling in a concept of fitness that is fundamentally teleological. In a famous passage, he discusses the transformation of a random sequence of letters into the Shakespearean quotation

> METHINKS IT IS LIKE A WEASEL

by a procedure that resembles a GA using mutation and selection. Such a transformation can be accomplished by random mutations, he argues, because when it gets a letter 'right', it is selected and need no longer be mutated. This is a feeble argument for *natural* selection (natural selection does not have a target, so how can it tell that it has a letter right without 'knowing' where it is 'going'?), and later Dawkins admits this, although he still appears to believe that his argument is sound. Without getting into a debate about the plausibility of his argument in its original context, it is certainly a useful illustration of how *artificial* selection could be used to solve problems.

Less polemical writers than Dawkins concede that we are still remarkably ignorant concerning the details of *a priori* fitness in the natural world—i.e., how we could measure it as a function of genotypic characteristics, and what form such a function might take. Natural organisms are noticeable for the occurrence both of pleiotropy (one gene has an effect on several phenotypic traits) and of polygeny (any one phenotypic characteristic is influenced by several genes). Mayr encapsulates the problem with a neat aphorism:

> *Genes are never soloists, they always play in an ensemble.* [159]

This characteristic is usually denoted by the use of the term *epistasis*: the fitness of the phenotype in a given environment is not the result of additive effects (either on an arithmetic or a logarithmic scale) at the genotypic level. In a recent survey Mitton asserts that in general

> *we do not yet understand how many loci interact to determine fitness or how selection will influence the genome's architecture.*[171]

Thus, to sum up, it is our contention that hand-waving references to neo-Darwinian evolution are insufficient to justify the use of GAs as a tool for optimization. A more limited but better analogy would be to the use of artificial selection in plant and animal breeding experiments. Here we have centuries of experience as evidence of the capacity of natural organisms to change in response to selective pressures imposed by some prior and external notion of fitness, albeit within very circumscribed limits in comparison to the all-embracing neo-Darwinian hypothesis.

However, even this does no more than demonstrate the potential of the idea of GA in solving optimization problems. That they can and do solve such problems is attested by many successful applications. However, to discover exactly how and why they work—to explore the GA's 'fitness' for the task—we need for the most part to put biology to one side, and to enter the world (and use the language) of mathematics. It is of interest that Holland, although working when neo-Darwinism had achieved what future generations may perceive as its high water mark, was not tempted to rely on purely analogical arguments. In his book [124], he provided an analysis of GA behaviour that he hoped would place GAs on a firm mathematical basis. Thus, by understanding the GA's *modus operandi* better, we would have a better idea of when and where to use them, and how they could be implemented to the best effect. (Perhaps, also, we would be able to place a few bricks in the foundations of a more comprehensive and substantial understanding of evolution itself.)

Initially, Holland's theoretical analysis seemed to provide exactly what was needed. However, in the last decade, evidence has accumulated that GAs do not necessarily behave in the way delineated by Holland, and we still do not have as straightforward an account as we would like. It is clear that GAs are fundamentally much more complicated than simulated annealing, for example. Over the last 10 years, many different ideas have been proposed to help us to understand how GAs work. In other words, we have seen the development of several different theoretical *perspectives* on GAs, none of which can claim to provide a complete answer, but which all help in their own way to illuminate the concept of a GA to some degree.

This book undertakes to describe some of these perspectives in as complete a way as possible, given the current state of knowledge, without overwhelming the reader with mathematical details. In Chapter 2, we review the basic principles of the genetic algorithm in some detail, as well as covering more advanced issues—operators, coding questions, population sizing and so on. GA novices will thus be given a thorough grounding in the subject, so that they will be well-equipped to implement their own algorithms. (And there is actually no substitute for 'trying it yourself', whether the goal is merely to solve a particular problem, or to enquire more deeply into the fundamental questions.)

In Chapter 3 we describe what may be called the 'traditional' theory of schemata developed by Holland, and popularized especially by Goldberg [96]. Before describing in more detail the different perspectives on GAs,

Chapter 4 provides a consideration of the implications of the 'no-free-lunch' theorem for GAs. This leads into Markov chain theory (Chapter 5), which enables the consideration of the dynamical behaviour of GAs (Chapter 6). While these approaches attempt to understand GAs at the micro-level, an opposite point of view is to focus on their macroscopic behaviour by using methods originally inspired by statistical mechanics; this is the subject of Chapter 7. Statistical methods are also at the heart of Chapters 8 and 9: firstly, we describe connections to experimental design methods in statistics that have been used in attempts to predict GA performance. This also relates to the idea of a landscape (Chapter 9), which enables us to link GAs to other heuristic search methods. In the concluding chapter we draw together what we can learn from these different approaches and discuss possible future directions in the development of GAs.

1.4 Bibliographic Notes

The 'classics' in evolutionary algorithms are the books by Lawrence Fogel [81, 82], John Holland [124, 126], Ingo Rechenberg [211] and Hans-Paul Schwefel [260]. David Goldberg's book [96] has been highly influential, while David Fogel's survey [79] is an important collection of historical documents.

Background material on heuristic search in optimization can be found in books by Aarts *et al.*[1], Papadimitriou and Steiglitz [185], and Reeves [213]. The handbook edited by Du and Pardalos [65] gives a more general picture of methods used to solve COPs. Two in-depth studies of the travelling salesman problem [148, 235] also provide considerable insight into combinatorial optimization in general.

The extent to which neo-Darwinian metaphors really apply to evolutionary algorithms has not often been considered: the paper by Daida *et al.*[43] is one notable exception that also contains an impressive bibliography of relevant writings. The literature on evolution itself is of course enormous: Darwin's own *magnum opus* [44] is available in many editions, but those who want a general modern introduction to evolution could try the concise and readable book by Patterson [186], or Ridley's much more detailed exposition [236]. The writings of Richard Dawkins—[51, 52, 53], for example—are very well known. Perhaps less well known is the extent to which Dawkins's view of evolution is controversial. In this, current thinking in Darwinist circles mirrors much of the 20th century: the disputes between (say) Dawkins and Stephen Jay Gould as depicted in [32] follow in a tradition exemplified by earlier conflicts between Fisher and Sewall Wright, and Fisher and Mayr.

Some of these aspects of the development of evolutionary thought are recounted by Depew and Weber [61], and Provine [193, 194]. Ernst Mayr himself provides his own unique historical perspective in [160].

Contemporary critics of neo-Darwinism who are worth reading for the purpose of stimulating thought include the philosopher Mary Midgley [168], and scientists such as James Shapiro [262] and Michael Behe [18], who specialize in biology at the molecular level, or those such as Brian Goodwin [103] and Stuart Kauffman [144, 145], who have a particular interest in issues to do with complex systems.

Chapter 2

Basic Principles

2.1 GA Concepts

We have already seen that Holland chose to use the language of biology, and described the basic structures that a GA manipulates as *chromosomes*. We will consider in the next section how such a structure might arise. For the moment, we will assume that we have a string or vector where each component is a symbol from an alphabet $\mathcal{A}$. A set[1] of such strings has been generated, and they collectively comprise the *population*. This population is about to undergo a process called *reproduction*, whereby the current population is transformed into a new one. This transformation takes place by means of *selection, recombination* and *mutation*. Numerous variants of GAs are possible, and we shall describe some of these later in the chapter.

2.2 Representation

As remarked in the previous chapter, the focus in this book is on using GAs as optimizers, so we recall here our basic optimization problem: given a function

$$g : \mathcal{V} \mapsto \mathbb{R},$$

find

$$\arg\min_{v \in \mathcal{V}} g.$$

From a mathematical perspective, we take this problem and encode the decision variables as strings or vectors where each component is a symbol

---

[1]To be strictly accurate, unless we take special measures to prevent it, the population is often a *multiset*, as multiple copies of one or more strings may be present.

from an alphabet $\mathcal{A}$. These strings are then concatenated to provide us with our chromosome. That is, the vector $\boldsymbol{v}$ is represented by a string $\boldsymbol{x}$ of symbols drawn from $\mathcal{A}$, using a mapping

$$c : \mathcal{A}^\ell \mapsto \mathcal{V}.$$

where ℓ, the length of the string, depends on the dimensions of both $\mathcal{V}$ and $\mathcal{A}$. The elements of the string correspond to 'genes', and the values those genes can take to 'alleles'. We shall call this the *encoding* function. This is also often designated as the *genotype-phenotype mapping*. The space $\mathcal{A}^\ell$—or, quite often, some subspace $\mathcal{X}$ of it—is known as the genotype space, while the space $\mathcal{V}$, representing the original definition of the problem, is the phenotype space.

Thus the optimization problem becomes one of finding

$$\min_{\boldsymbol{x} \in \mathcal{X}} g(\boldsymbol{x})$$

where (with a slight abuse of notation)

$$g(\boldsymbol{x}) = g(c(\boldsymbol{x})).$$

In practice we often blur the distinction between genotype and phenotype and refer to $g(\boldsymbol{v})$ or $g(\boldsymbol{x})$ according to context. Often also we make use of a monotonic transformation of g to provide what is usually termed a *fitness function*, which we can regard as another function

$$f : \mathcal{X} \mapsto \mathbb{R}^+.$$

This has the merit of ensuring that fitness is always a positive value. We note also that the genotype search space

$$\mathcal{X} \subseteq \mathcal{A}^\ell,$$

which implies that some strings may represent invalid solutions to the original problem. This can be a source of difficulty for GAs, since GA operators are not guaranteed to generate new strings in $\mathcal{X}$.

The choice of encoding is usually dictated by the type of problem that is being investigated. We shall now describe some of the more common situations.

2.2. REPRESENTATION

2.2.1 Binary problems

In some problems a binary encoding might arise naturally. Consider the operational research problem known as the *knapsack* problem, stated as follows.

Example 2.1 (The 0-1 knapsack problem) *A set of n items is available to be packed into a knapsack with capacity C units. Item i has a weight (or value) w_i and uses up c_i units of capacity. The objective is to determine the subset I of items to pack in order to maximize*

$$\sum_{i \in I} w_i$$

such that

$$\sum_{i \in I} c_i \leq C.$$

If we define

$$x_i = \begin{cases} 1 & \text{if item } i \text{ is packed} \\ 0 & \text{otherwise} \end{cases}$$

the knapsack problem can be re-formulated as an *integer program*:

$$maximize \sum_{i=1}^{n} w_i x_i$$

$$such\ that \sum_{i=1}^{n} c_i x_i \leq C$$

from which it is clear that we can define a solution as a binary string of length n. In this case there is thus no distinction between genotype and phenotype. However, not all binary strings are feasible solutions, and it may be difficult to ensure that the genetic operators do not generate solutions outside the range of $\mathcal{X}$.

2.2.2 Discrete non-binary problems

There are cases in which a discrete alphabet of higher cardinality might be appropriate. The rotor stacking problem, as originally described by McKee and Reed [162], is a good example.

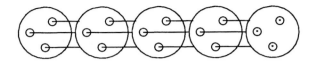

Figure 2.1: Rotor stacking problem.

Example 2.2 *A set of n rotors is available, each of which has q holes drilled in it. The rotors have to be assembled into a unit by stacking them and bolting them together, as in Figure 2.1. Because the rotors are not perfectly flat, the assembly will show deviations from true symmetry, with the consequent effect that in operation the assembled unit will wobble as it spins. The objective is to find which of all the possible combinations of orientations produces the least deviation.*

In this case a q-ary coding is natural. A solution is represented by a string of length n, each gene corresponding to a rotor and the alleles, drawn from $\{1, \ldots, q\}$, representing the orientation (relative to a fixed datum) of the holes. Thus, the string (1 3 2) represents a solution to a 3-rotor problem where hole 1 of the first rotor is aligned with hole 3 of the second and hole 2 of the third. Of course, it would be possible to encode the alleles as binary strings, but there seems little point in so doing—particularly if q is not a power of 2, as some binary strings would not correspond to any actual orientation.

This seems very straightforward, but there is a subtle point here that could be overlooked. The assignment of labels to the holes is arbitrary, and this creates a problem of competing conventions. This can be alleviated in this case by fixing the labelling for one rotor, so that a solution can be encoded by a string of length $(n-1)$.

2.2.3 Permutation problems

There are also some problems where the 'obvious' choice of representation is defined, not over a set, but over a permutation. As an example, consider the permutation flowshop sequencing problem (PFSP).

2.2. REPRESENTATION

Example 2.3 *Suppose we have n jobs to be processed on m machines, where the processing time for job i on machine j is given by $p(i,j)$. For a job permutation $\pi = (\pi_1, \pi_2, \cdots, \pi_n)$, we calculate the completion times $C(\pi_i, j)$ as follows:*

$$\begin{aligned}
C(\pi_1, 1) &= p(\pi_1, 1) \\
C(\pi_i, 1) &= C(\pi_{i-1}, 1) + p(\pi_i, 1) \quad \text{for } i = 2, \ldots, n \\
C(\pi_1, j) &= C(\pi_1, j-1) + p(\pi_1, j) \quad \text{for } j = 2, \ldots, m \\
C(\pi_i, j) &= \max\{C(\pi_{i-1}, j), C(\pi_i, j-1)\} + p(\pi_i, j) \\
&\quad \text{for } i = 2, \ldots, n; \; j = 2, \ldots, m
\end{aligned}$$

A common objective relates to the makespan

$$C_{max}(\pi) = C(\pi_n, m).$$

The makespan version of the PFSP is to find a permutation $\pi^* \in \Pi_n$ such that

$$C_{max}(\pi^*) \leq C_{max}(\pi) \; \forall \pi \in \Pi_n.$$

Here the natural encoding (although not the only one) is simply the permutation of the jobs as used to calculate the completion times. So the solution (1 4 6 2 5 3 7), for example, simply means that job 1 is first on each machine, then job 4, job 6, etc.

Unfortunately, the standard crossover operators fail to preserve the permutation except in very fortunate circumstances, and specific techniques have had to be devised for this type of problem. Some of these will be discussed later in Section 2.3.5.

2.2.4 Non-binary problems

In many cases the natural variables for the problem are not binary, but integer or real-valued. In such cases a transformation to a binary string is required first. Note that this is a different situation from the rotor-stacking example, where the integers were merely labels: here the values are assumed to be meaningful as numbers.

Example 2.4 *It is required to maximize*

$$f(x) = x^3 - 60x^2 + 900x + 100$$

over the search space $\mathcal{V} = \{x : x \in \mathbb{Z}; x \in \{0,31\}\}$, i.e., the solution x^* is required to be an integer in the range $[0, 31]$.

To use the conventional form of genetic algorithm here, we would use a string of 5 binary digits with the standard binary to integer mapping, i.e., the encoding function $c(\cdot)$ in this case is such that

$$\begin{cases} 0\,0\,0\,0\,0 & = & 0 \\ 0\,0\,0\,0\,1 & = & 1 \\ \cdots\cdots\cdots\cdots \\ 1\,1\,1\,1\,1 & = & 31 \end{cases}$$

Of course, in practice we could solve such a problem easily without recourse to encoding the decision variable in this way, but it illustrates neatly the sort of optimization problem to which GAs are often applied. Such problems assume firstly that we know the domain of each of our decision variables, and secondly that we have some idea of the precision with which we need to specify our eventual solution. Given these two ingredients, we can determine the number of bits needed for each decision variable, and concatenate them to form the chromosome. In the general case, if a decision variable can take values between a and $b(> a)$, and it is mapped to a binary string of length ℓ, the precision is

$$\frac{b-a}{2^\ell - 1}.$$

Assuming we know what precision is required, this can be used to find a suitable value for ℓ. Note that it may differ from one decision variable to another, depending on the degree of precision required, and on prior knowledge of the range of permissible values. Of course, in many cases, this prior knowledge may be rather vague, so that the range of values is extremely wide. A helpful technique in such cases is *dynamic parameter encoding* [258], where the range of values—initially large—is allowed to shrink as the population converges. A similar technique is *delta coding* [303], in which successive trials are made in which at each iteration the parameters encoded are offsets from the best solution found at the previous iteration. Both techniques thus proceed from an initial coarse-grained search through phases of increasing refinement.

It should also be pointed out that the conventional binary-to-integer mapping is not the only one possible, nor is it necessarily the best. Some research suggests that Gray codes are often preferable. Using the standard binary code, adjacent genotypes may have distant phenotypes, and *vice-versa*, whereas Gray codes do at least ensure that adjacent phenotypes always correspond to adjacent genotypes. More will be said on Gray codes in Chapters 4 and 9.

Finally, in cases like these where the variables are continuous, some practitioners would argue that the use of an evolution strategy (ES) is preferable, and that binary encoding adds an unnecessary layer of confusion. We shall not enter this debate here, but interested readers will find helpful advice and insights in the book by Bäck [12].

2.3 The Elements of a Genetic Algorithm

A basic version of a GA works as shown in Figure 2.2. Even this is sufficient to show that a GA is significantly more complicated than neighbourhood search methods, with several interacting elements. These various elements will now be described in more detail.

2.3.1 Initial population

The major questions to consider are firstly the size of the population, and secondly the method by which the individuals are chosen. The choice of the population size has been approached from several theoretical points of view, although the underlying idea is always of a trade-off between efficiency and effectiveness. Intuitively, it would seem that there should be some 'optimal' value for a given string length, on the grounds that too small a population would not allow sufficient room for exploring the search space effectively, while too large a population would so impair the efficiency of the method that no solution could be expected in a reasonable amount of computation. Goldberg [93] was probably the first to attempt to answer this question, using the idea of schemata, as outlined in the next chapter. Unfortunately, from this viewpoint, it appeared that the population size should increase as an exponential function of the string length. Some refinements of this work are reported in [97], but they do not change the overall conclusions significantly.

Fortunately, empirical results from many authors (see e.g., Grefenstette [106] or Schaffer *et al.* [254]) suggest that population sizes as small as 30 are quite adequate in many cases. A later analysis from a different perspective [101] led Goldberg and colleagues to the view that a linear dependence of population size on string length was adequate. That populations should

```
Choose an initial population of chromosomes;
while termination condition not satisfied do
    repeat
        if crossover condition satisfied then
        {select parent chromosomes;
        choose crossover parameters;
        perform crossover};
        if mutation condition satisfied then
        {select chromosome(s) for mutation;
        {choose mutation points;
        perform mutation};
        evaluate fitness of offspring
    until sufficient offspring created;
    select new population;
endwhile
```

Figure 2.2: A genetic algorithm template. This is a fairly general formulation, accommodating many different forms of selection, crossover and mutation. It assumes user-specified conditions under which crossover and mutation are performed, a new population is created, and whereby the whole process is terminated.

grow with string length is perfectly plausible, but even a linear growth rate would lead to quite large populations in some cases.

A slightly different question that we could ask is regarding a *minimum* population size for a meaningful search to take place. In [213], the principle was adopted that, at the very least, every point in the search space should be reachable from the initial population by crossover only. This requirement can only be satisfied if there is at least one instance of every allele at each locus in the whole population of strings. On the assumption that the initial population is generated by a random sample with replacement (which is a conservative assumption in this context), the probability that at least one allele is present at each locus can be found. For binary strings this is easily seen to be

$$P_2^* = (1 - (1/2)^{N-1})^\ell.$$

Using an exponential function approximation

$$P_2^* \approx \exp\left(-\ell/2^{N-1}\right)$$

we can easily establish that

$$N \approx \lceil 1 + \log\left(-\ell/\ln P_2^*\right)/\log 2 \rceil.$$

For example, a population of size 17 is enough to ensure that the required probability exceeds 99.9% for strings of length 50. For q-ary alphabets (where $q > 2$), the calculation is somewhat less straightforward, but [213] derives expressions that can be converted numerically into graphs for specified confidence levels. Figure 2.3 gives an example for 99.9% confidence and alphabets of different cardinalities.

While this is not quite the same as finding an optimal size, it does help to account for the good reported performance of GAs with populations as low as 30 for a variety of binary-encoded problems. It can be seen that in a randomly generated population of binary strings we are virtually certain to have every allele present for populations of this size. Furthermore, the growth rate with string length appears to be $\mathcal{O}(\log \ell)$, which is better (i.e., slower) than linear for all values of q.[2]

This analysis also shows that we need disproportionately larger populations for higher-cardinality alphabets. We can compare, for example the cases $q = 2$ with $q = 8 = 2^3$. In the second case, we would have a string 3

---

[2] Treating the step functions in Figure 2.3 as curves and fitting $N = a + b \log \ell$ gives a coefficient of determination $R^2 \approx 100\%$ to several places of decimals. Both a and b also appear to increase slowly with q, but in a linear fashion.

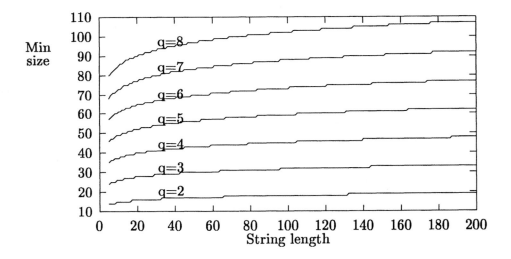

Figure 2.3: Minimal population sizes for 99.9% confidence that all alleles are present in the initial population.

2.3. THE ELEMENTS OF A GENETIC ALGORITHM

times as long if we were to use a binary encoding of the 8 symbols in the larger alphabet. However, the minimal population size for an 8-ary string of length ℓ is about 6 times that needed for a binary string of length 3ℓ. While there is often a 'natural' encoding for a given problem, this suggests that, other things being equal, the computational burden is likely to be greater for a higher-cardinality alphabet. Of course, whether other things *are* equal is a matter of some debate.

In respect of how the initial population is chosen, it is nearly always assumed that initialization should be 'random', which in practice means the use of pseudorandom sequences. However, points chosen in this way do not necessarily cover the search space uniformly, and there may be advantages in terms of coverage if we use more sophisticated statistical methods, especially if the alphabet is non-binary. One relatively simple idea is a generalization of the Latin hypercube [161]. Rather than rely on chance to generate all the alleles we need, we could insert them directly into the population. This idea can be illustrated as follows.

Suppose each gene has 5 alleles, labelled $0, \ldots, 4$. We choose the population size N to be a multiple of 5, and the alleles in each 'column' are generated as an independent random permutation of $0, \ldots, N-1$, which is then taken modulo 5. Figure 2.4 shows an example for a population of size 10. To obtain search space coverage at this level with simple random initialization would need a much larger population. Some other possibilities for GAs are discussed by Kallel and Schoenauer [139]. More widely, it should be noted that in OR the value of using 'low discrepancy sequences' instead of pseudorandom ones is being increasingly recognized. L'Ecuyer and Lemieux [149] give a recent survey; whether such techniques have a place in evolutionary algorithms is so far unexplored.

A final point to mention here is the possibility of seeding the initial population with known good solutions. Some reports (for example, in [4, 216]) have found that including a high-quality solution, obtained from another heuristic technique, can help a GA find better solutions rather more quickly than it can from a random start. However, there is also the possibility of inducing premature convergence [143, 151] to a poor solution. Surry and Radcliffe [275] review a number of ideas related to the question of population initialization under the heading of *inoculation*, and also conclude that, relative to completely random initialization, seeding tends to reduce the quality of the best solution found.

Individual	Gene	Gene
	1 2 3 4 5 6	1 2 3 4 5 6
1	0 1 3 0 2 4	0 0 1 3 0 3
2	1 4 4 2 3 0	1 3 4 0 4 4
3	0 0 1 2 4 3	3 2 3 1 2 4
4	2 4 0 3 1 4	2 3 3 3 0 1
5	3 3 0 4 4 2	0 0 2 4 0 0
6	4 1 2 4 3 0	1 0 4 0 4 4
7	2 0 1 3 0 1	1 4 2 2 1 0
8	1 3 3 1 2 2	2 0 2 0 0 2
9	4 2 2 1 1 3	3 1 0 3 4 0
10	3 2 4 0 0 1	1 4 2 2 1 3

Figure 2.4: On the left is an example of Latin hypercube sampling for $\ell = 6$ and $|\mathcal{A}| = 5$. Notice that each allele occurs exactly twice for each gene. In contrast, on the right is a typical randomly selected population, where in two cases (genes 1 and 5) an allele (4,3 respectively) never occurs at all.

2.3.2 Termination

Unlike simple neighbourhood search methods that terminate when a local optimum is reached, GAs are stochastic search methods that could in principle run for ever. In practice, a termination criterion is needed; common approaches are to set a limit on the number of fitness evaluations or the computer clock time, or to track the population's diversity and stop when this falls below a preset threshold. The meaning of diversity in the latter case is not always obvious, and it could relate either to the genotypes or the phenotypes, or even, conceivably, to the fitnesses, but in any event we need to measure it by statistical means. For example, we could decide to terminate a run if at every locus the proportion of one particular allele rose above 90%.

2.3.3 Crossover condition

Given the stress on recombination in Holland's original work, it might be thought that crossover should always be used, but in fact there is no reason to suppose that it has to be so. Thus, while we could follow a strategy of crossover-AND-mutation to generate new offspring, it is also possible to use crossover-OR-mutation. There are many examples of both in the literature. The first strategy carries out crossover (possibly with probability less than 1), then attempts mutation on the offspring (either or both). It is conceivable that in some cases nothing happens at all with this strategy—neither crossover or mutation is actually applied, so that the offspring are simply clones of the parents. In the second case, the implementation prefer always to do something, either crossover of mutation, but not both.

The mechanism for implementing such probabilistic choices is normally a randomized rule, whereby the operation is carried out if a pseudo-random uniform deviate falls below a threshold value. In the case of crossover, this is often called the crossover rate, and we shall later use the symbol χ to denote this value. For mutation, we have a choice between describing the number of mutations per string, or per gene; in this book we shall use the gene-wise mutation rate, and denote it by the symbol μ.

In the -OR- case, there is the further possibility of modifying the relative proportions of crossover and mutation as the search progresses. Davis [47] has argued that different rates are appropriate at different times: high crossover at the start, high mutation as the population converges. In fact, he has suggested that the operator proportions could be adapted online, in accordance with their track record in finding new high-quality chromosomes.

2.3.4 Selection

The basic idea of selection is that it should be related to fitness, and the original scheme for its implementation is commonly known as the *roulette-wheel* method. Roulette wheel selection (RWS) uses a probability distribution in which the selection probability of a given string is directly proportional to its fitness. Figure 2.5 provides a simple example of roulette-wheel selection. Pseudo-random numbers are used one at a time to choose strings for parenthood. For example, in Figure 2.5, the number 0.135 would select string 1, the number 0.678 would select string 4.

Finding the appropriate number for a given pseudo-random number r requires searching an array for values that bracket r—this can be done using a binary search in $\mathcal{O}(\log N)$ time. However, this approach has a high stochastic variability, and the actual number of times N_C that chromosome C is selected in any generation may be very different from its expected value $E[N_C]$. For this reason, sampling *without* replacement may be used, to ensure that at least the integral part of $E[N_C]$ is achieved, with fractions being allocated using random sampling.

In practice, Baker's *stochastic universal selection* (SUS) [15] is a particularly effective way of realizing this outcome. Instead of a single choice at

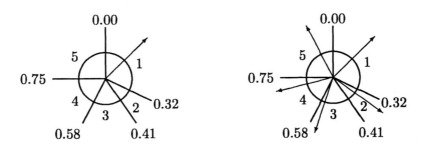

Figure 2.5: Suppose there are 5 strings in a population with fitnesses 32, 9, 17, 17, 25 respectively. The probability of selection of each individual is proportional to the area of a sector of a roulette wheel (or equivalently, to the angle subtended at the centre). The numbers on the spokes of the wheel are the cumulative probabilities for use by a pseudo-random number generator. On the left we have standard roulette-wheel selection, with a single pointer that has to be spun 5 times. On the right we have SUS, using 5 connected equally-spaced pointers; one spin provides 5 selections.

each stage, we imagine that the roulette wheel has an equally spaced multi-armed spinner. Spinning the wheel produces simultaneously the values N_C for all the chromosomes in the population. From the viewpoint of statistical sampling theory, this corresponds to systematic random sampling [153]. Experimental work by Hancock [111] clearly demonstrates the superiority of this approach, although much published work on applications of GAs still unaccountably appears to rely on the basic roulette-wheel method. This paper, and an updated version in [112], give a thorough experimental analysis of several different selection methods.

Finally, we may need to define what happens if we inadvertently try to mate a string with itself. For reasonably large populations and string lengths, this is unlikely to be a problem in the early stages. However, as the population converges, attempts to mate a string with itself or a clone may not be uncommon. This could be taken as a signal for termination, as discussed in the previous section. Alternatively, a 'no duplicates' policy could be operated to prevent it from happening at all.

2.3. THE ELEMENTS OF A GENETIC ALGORITHM

Scaling

An associated problem is that of finding a suitable measure of fitness for the members of the population. Simply using the objective $g(x)$ is rarely sufficient, because the scale on which $g(x)$ is measured is important. (For example, values of 10 and 20 are much more clearly distinguished than 1010 and 1020.) Further, if the objective is minimization rather than maximization, a transformation is clearly required.

Some sort of scaling is thus often applied, and Goldberg [96] gives a simple algorithm to deal with both minimization and maximization. A simple linear transformation is used to convert from the objective g to the fitness f, i.e.,

$$f = ag + b$$

where the parameters a and b are obtained from two conditions: firstly, that the means should be the same

$$\bar{f} = \bar{g},$$

and secondly, that the maximum fitness is a constant multiple of its mean:

$$f_{max} = \phi \bar{f}$$

for some scalar ϕ. The method is cumbersome, however, and continual rescaling is needed as the search progresses. Two alternatives provide more elegant solutions.

Ranking

Ranking the chromosomes in fitness order loses some information, but there is no need for re-scaling, and selection algorithm is simpler and more efficient. Suppose the probability of selecting the string that is ranked kth in the population is denoted by $\mathbf{Pr}[k]$. In the case of linear ranking, we assume that

$$\mathbf{Pr}[k] = \alpha + \beta k$$

where α and β are positive scalars. The requirement that $\mathbf{Pr}[k]$ be a probability distribution gives us one condition:

$$\sum_{k=1}^{N}(\alpha + \beta k) = 1 \Rightarrow N(\alpha + \beta \frac{N+1}{2}) = 1$$

which leaves us free to choose the other parameter in a way that tunes the *selection pressure*. This term is loosely used in many papers and articles on GAs. In this book, we mean the following:

Definition 2.1 *Selection pressure*

$$\phi = \frac{\Pr[\text{selecting fittest string}]}{\Pr[\text{selecting average string}]}.$$

(Note that this formalises the parameter ϕ in Goldberg's scaling approach.) In the case of linear ranking, we interpret the average as meaning the *median* string, so that

$$\phi = \frac{\alpha + \beta N}{\alpha + \beta(N+1)/2}.$$

Some simple algebra soon establishes that

$$\beta = \frac{2(\phi - 1)}{N(N-1)} \quad \text{and} \quad \alpha = \frac{2N - \phi(N+1)}{N(N-1)}$$

which implies that $1 \leq \phi \leq 2$. Within this framework, it is easy to see that the cumulative probability distribution can be stated in terms of the sum of an arithmetic progression, so that finding the appropriate k for a given pseudo-random number r is simply a matter of solving the quadratic equation

$$\alpha k + \beta \frac{k(k+1)}{2} = r$$

for k, which can be done simply in $\mathcal{O}(1)$ time. The formula is

$$k = \frac{-(2\alpha + \beta) + \sqrt{(2\alpha + \beta)^2 + 8\beta r}}{2\beta}.$$

In contrast, searching for k for a given r using ordinary fitness-proportional selection needs $\mathcal{O}(\log N)$ time. However, finding the initial ranks requires a sorting algorithm, which at best still needs $\mathcal{O}(N \log N)$ for each new generation, so the gain in computational efficiency is illusory. Nevertheless, the possibility of maintaining a constant selection pressure without the need for re-scaling is an attractive one.

Other ranking functions (such as geometrically increasing probabilities) can be used besides linear ranking [301] but the above scheme is sufficiently flexible for most applications.

Tournament selection

The other alternative to strict fitness-proportional selection is *tournament selection* in which a set of τ chromosomes is chosen and compared, the best one being selected for parenthood. This approach has similar properties

2.3. THE ELEMENTS OF A GENETIC ALGORITHM

to linear ranking. In a complete cycle (generating N new chromosomes), each chromosome will be compared τ times on average, and it is easy to see that the best string will be selected every time it is compared. However, the chance of the median string being chosen is the probability that the remaining $\tau - 1$ strings in its set are all worse:

$$\left(\frac{1}{2}\right)^{\tau-1},$$

so that the selection pressure (as defined earlier) is

$$\phi = 2^{\tau-1}$$

if we assume a *strict* tournament; i.e., one in which the best string always wins. To obtain a selection pressures below $\phi = 2$, *soft* tournaments can be used—the best string is only given a chance $p < 1$ of winning. In this case the selection pressure is

$$\phi = 2p.$$

The properties of tournament selection are very similar to those of ranking. It is not hard to show that using a soft tournament with $\tau = 2$ produces the same selection pressure (in fact, the same probabilities) as linear ranking. Similarly, as shown by Julstrom [138], geometric ranking with probabilities

$$\mathbf{Pr}[k] = \alpha \beta^k$$

is approximately the same as a strict tournament if

$$\beta = \left(\frac{N}{N-1}\right)^\tau.$$

One potential advantage of tournament selection over all other forms is that it only needs a preference ordering between pairs or groups of strings, and it can thus cope with situations where there is no formal objective function at all—in other words, it can deal with a purely *subjective* objective function! It is also useful in cases where fitness evaluation is expensive; it may be sufficient just to carry out a partial evaluation in order to determine the winner.

However, we should point out again that tournament selection is also subject to arbitrary stochastic effects in the same way as roulette-wheel selection—there is no guarantee that every string will appear in a given population selection cycle. Indeed, using sampling with replacement there is a probability of about $e^{-1} (\approx 0.368)$ that a given string will not appear at all during a selection cycle. One way of coping with this, at the expense of a little extra computation, is to use a variance reduction technique from simulation theory. Saliby [250] distinguishes between the *set* effect and the *sequence* effect in drawing items from a finite population. In applying his ideas here, we know that we need τ items to be drawn N times, so we simply construct τ random permutations[3] of the numbers $1, \ldots, N$—the indices of the individuals in the population. These are concatenated into one long sequence which is then chopped up into N pieces, each containing the τ indices of the individuals to be used in the consecutive tournaments. Figure 2.6 gives an example. If N is not an exact multiple of τ, there is the small chance of some distortion where the permutations 'join', but this is not hard to deal with.

Three permutations of $\{1,2,3,4\}$		
1 4 2 3	2 1 3 4	1 2 4 3
The resulting four tournaments			
1 4 2	3 2 1	3 4 1	2 4 3

Figure 2.6: Tournament selection for the case $N = 4, \tau = 3$.

Selection intensity

As defined above, selection pressure is directly related to the user's choice of selection strategy and parameters. Some researchers have used a different means of characterizing selection—*selection intensity*. This concept originated in population genetics with experimental work for artificial breeding of plants or animals, and it was introduced to the GA community by Mühlenbein and Schlierkamp-Voosen [175], whose 'breeder GA' will be discussed in more detail later in Section 2.6.1. Selection intensity is defined as

$$I(t) = \frac{\bar{f}_{sel}(t) - \bar{f}(t)}{\sigma_f(t)},$$

i.e., the difference between the mean fitness of those chromosomes selected for reproduction and the mean fitness of the population, scaled by the population standard deviation. In animal breeding, yet another selection scheme is customarily used: *truncation selection*. In this approach only strings in

---

[3]There is a simple algorithm for doing this efficiently—see Nijenhuis and Wilf[181], for example, or Skiena's web repository [266].

2.3. THE ELEMENTS OF A GENETIC ALGORITHM

the top $P\%$ of the population are eligible to reproduce; within this group parents are chosen uniformly at random.

Selection intensity is clearly an *a posteriori* measure, and one that is also specific to the current population; if we make some assumption about the population fitness distribution (commonly, that the distribution is stationary and Normal), the value becomes a constant $I(t) = I$. Nevertheless, the value of I will still depend on the control parameters used for selection in a non-trivial way. While it is conceptually a more useful basis for comparing different GAs, it is much more difficult to estimate than the simple selection pressure. Further, it really only applies to strictly generational GAs and not to the élitist and steady-state versions, which will be described shortly in Section 2.3.7.

Blickle and Thiele [24] have surveyed several popular GA selection schemes and compared their theoretical values of I for a Normal fitness distribution. Their results enable the calculation of selection intensities from the control parameters for these selection schemes, and thus to determine parameter settings for different schemes that give the same (expected) selection intensities. They also calculated some 'second-order' properties involving measures of the 'selection variance' and 'loss of diversity'. A notable result was that truncation selection involves a higher loss of diversity than any other method tested. Thierens [278] has recently extended this work to the case of non-generational GAs.

Takeover time

Another measure sometimes used to compare selection schemes is the *takeover time*. This computes the time taken for the best string to take over the whole population, assuming only selection and no crossover or mutation. It depends on the population size as well as the parameters used for implementing selection. Hancock's experimental work [112], referred to above, concentrated on takeover time as a measure.

Theoretical results can also be obtained. Goldberg and Deb [100] make some simplifying assumptions for several selection strategies in order to obtain takeover time curves, which they found usually agreed well with experimentally determined values. Their main finding was that takeover time tended to grow as $\mathcal{O}(\log N)$ for most strategies except fitness-proportional selection, whose growth curve was $\mathcal{O}(N \log N)$. They also gave a comparison of time complexity, which showed that SUS and tournament selection are the cheapest in terms of computational requirements. Rudolph [248, 249] has analysed a variety of more complicated selection schemes, all of which turn out to have an $\mathcal{O}(N \log N)$ growth for takeover times.

2.3.5 Crossover

Having selected parents in some way, we need to recombine them—the process called *crossover*. Crossover is simply a matter of replacing some of the alleles in one parent by alleles of the corresponding genes of the other. Suppose that we have 2 strings a and b, each consisting of 7 variables, i.e.

$$(a_1, a_2, a_3, a_4, a_5, a_6, a_7) \quad \text{and} \quad (b_1, b_2, b_3, b_4, b_5, b_6, b_7),$$

which represent two possible solutions to a problem. To implement *one-point crossover* (denoted by 1X), a *crossover point* is chosen at random from the numbers $1, \ldots, 6$, and a new solution produced by combining the pieces of the original 'parents'. For instance, if the crossover point was 2, then the 'offspring' solutions would be

$$(a_1, a_2, b_3, b_4, b_5, b_6, b_7) \quad \text{and} \quad (b_1, b_2, a_3, a_4, a_5, a_6, a_7).$$

A similar prescription can be given for m-point crossover where $m > 1$.

An early and thorough investigation of multi-point crossovers is that by Eshelman *et al.* [72], who examined the biasing effect of traditional one-point crossover, and considered a range of alternatives. Their central argument is that two sources of bias exist to be exploited in a genetic algorithm: *positional* bias, and *distributional* bias. Simple crossover (1X) has considerable positional bias, in that it favours substrings of contiguous bits, and we cannot be sure that this bias is working in favour of good solutions.

On the other hand, simple crossover has no distributional bias, in that the crossover point is chosen randomly using the uniform distribution. But this lack of bias is not necessarily a good thing either, as it limits the exchange of information between the parents. In [72], the possibilities of changing these biases, in particular by using multi-point crossover, were investigated and empirical evidence strongly supported the suspicion that one-point crossover is not the best option. Although the evidence was somewhat ambiguous, it seemed to point to an 8-point crossover operator as the best overall, in terms of the number of function evaluations needed to reach the global optimum.

Another obvious alternative, which removes any bias, is to make the crossover process completely random—the so-called uniform crossover. This can be seen most easily by observing that the crossover operator itself can be written as a binary string or *mask*—in fact, when implementing any

2.3. THE ELEMENTS OF A GENETIC ALGORITHM

crossover in a computer algorithm, this is the obvious way to do it. Any linear crossover operator γ can be represented by a vector $m \in \{0,1\}^\ell$ so that the offspring of parents a and b is

$$m \otimes a \oplus \overline{m} \otimes b$$

where $\overline{m}$ is the complement of m, and $\oplus, \otimes$ denote component-wise addition and multiplication respectively. For example, the mask

$$1100011$$

represents a 2-point crossover (2X), the offspring of a and b (as defined previously) being the vector $(a_1, a_2, b_3, b_4, b_5, a_6, a_7)$. (Another offspring can be obtained by switching the rôles of a and b or, alternatively, by switching the rôles of m and $\overline{m}$.) By generating the pattern of 0s and 1s stochastically (using a Bernoulli distribution) we get uniform crossover (UX), which might generate a mask such as

$$1010001$$

This idea was first used by Syswerda [277], who implicitly assumed that the Bernoulli parameter $p = 0.5$. Of course, this is not necessary: we can bias UX towards one or other of the parents by choosing p appropriately.

De Jong and Spears [57] produced a theoretical analysis that was able to characterize the amount of disruption introduced by a given crossover operator exactly. In particular, the amount of disruption in UX can be tuned by choosing different values of p.

Non-linear crossover

Treating crossover as a linear operator makes sense when dealing with binary strings. However, when the representation is non-linear, crossover has to be reinterpreted. One of the most frequently occurring problems is where the solution space is the space of permutations Π_ℓ—well-known examples of this include many scheduling problems, and the famous travelling salesman problem (TSP).

For instance, the simple-minded application of 1X in the following case produces an infeasible solution:

P1	1 6 3 4 7 5 2	O1	1 6 1 2 6 5 7
	X		
P2	4 3 1 2 6 5 7	O2	4 3 3 4 7 5 2

If this represents a TSP, the first offspring visits cities 1 and 6 twice, and never reaches cities 3 or 4 at all, these problems being mirrored in the second offspring's tour. A moment's thought is enough to realize that this type of behaviour will be the rule, not an exception. Clearly we need to think of something rather smarter if we are to be able to solve such problems.

One of the first ideas for such problems was the PMX (partially matched crossover) operator [94]. Two crossover points are chosen uniformly at random between 1 and $\ell - 1$. The section between these points defines an interchange mapping. Thus PMX might proceed as follows:

```
P1   1 6 3 4 7 5 2    O1   3 7 1 2 6 5 4
       X   X
P2   4 3 1 2 6 5 7    O2   2 1 3 4 7 5 6
```

Here the segment between the crossover points defines an interchange mapping

$$3 \leftrightarrow 1; \quad 4 \leftrightarrow 2; \quad 7 \leftrightarrow 6$$

on their respective strings, which means that the cut blocks have been swapped and now appear in different contexts from before. Another possibility is to apply a binary mask, as in linear crossover, but with a different meaning. Such a mask, generated as with UX, say, might be the following

$$1 0 1 0 0 1 0$$

which is applied to the parents in turn. First the components corresponding to 1s are copied from one parent, and then those that correspond to 0s are taken in the order they appear from the second parent in order to fill the gaps. Thus the above example generates the following pairs of strings:

```
P1   1 6 3 4 7 5 2   ⇒   1 - 3 - - 5 -   ⇒   O1   1 4 3 2 6 5 7
P2   4 3 1 2 6 5 7   ⇒   4 - 1 - - 5 -   ⇒   O2   4 6 1 3 7 5 2
```

This is one version of what is sometimes called *order* crossover, but it by no means exhausts the possibilities for dealing with permutations. A comprehensive discussion of possible methods of attack is contained in papers by Fox and McMahon [85] and Poon and Carter [190]. Furthermore, particular problems raise particular issues. In the case of the TSP, for example, the *positions* of the cities in the string is less important than their adjacencies, as has long been realized in OR when implementing local search methods for the TSP. Recognising this, Whitley *et al.* [304] proposed the *edge recombination* crossover for the TSP. Recombination becomes a matter of selecting

2.3. THE ELEMENTS OF A GENETIC ALGORITHM

edges from the combined edge sets of the two parents, rather than selecting city positions from two permutations. More recent work by Nagata and Kobayashi [177] has also concentrated on edges—their *edge assembly crossover* was the first case where GAs reliably found optimal solutions for large (> 3000 cities) TSPs.

Generalized n-point crossover

In fact, the TSP is only one of a substantial class of problems for which subset selection is the key rather than linear combinations of strings. Radcliffe has analysed such problems very thoroughly in a series of papers [198, 199, 200, 202], and the book by Falkenauer [76] also covers some examples of this type.

The problem of needing different forms of crossover for different representations led Radcliffe and Surry [204] to propose a generalized form of crossover (GNX) suitable for any representation. It is best described by the template in Figure 2.7.

Choose a random integer $n \in [1, \ell - 1]$;
Choose n crosspoints;
Generate a random permutation σ of $(1, \ldots, n+1)$ for segment order;
Designate one parent for copying;
$k \leftarrow 1$;
   repeat
      copy all compatible alleles of segment σ_k from designated parent;
      swap parent designations
      $k \leftarrow k + 1$;
   until $k = n + 1$;
if child incomplete **then** insert legal allele(s) at required position, using random tie-breaking if necessary.

Figure 2.7: A template for generalized n-point crossover. Parental rôles are swapped at successive iterations and only compatible alleles are copied.

For the example above, suppose $n = 2$, with crosspoints at loci 2 and 5, and the segment order is $\sigma = (1, 3, 2)$.

$$\begin{array}{ll} \text{P1} & 1\,6\,3\,4\,7\,5\,2 \\ & \text{X}\text{X} \\ \text{P2} & 4\,3\,1\,2\,6\,5\,7 \end{array}$$

The designated parent for the assignment of the first segment (segment 1) is P1, which gives

$$1\,6\,-\,-\,-\,-\,-\,;$$

for the second (segment 3) it is P2, giving

$$1\,6\,-\,-\,-\,5\,7\,;$$

and for the last segment (segment 2) it is P1 again, so we obtain

$$1\,6\,3\,4\,-\,5\,7$$

Finally, allele 2 is the only legal assignment possible, so the offspring is

$$1\,6\,3\,4\,2\,5\,7\,.$$

The definition of 'compatible' or 'legal' alleles may still be problem-specific, so this may appear over-complicated and unnecessary, and of course, standard versions of crossover emerge as special cases. However, the advantage is that the same framework will apply should the alleles be edges instead of cities. Using this edge set representation means we merely have two *sets* of edges, which could be placed in any order at all. However, for clarity of comparison we will maintain the order used above for the 'allele=city' representation.

$$\begin{array}{cc} \text{P1} & \{1\text{-}6,6\text{-}3,3\text{-}4,4\text{-}7,7\text{-}5,5\text{-}2,2\text{-}1\} \\ & \text{X} \qquad\qquad\qquad \text{X} \\ \text{P2} & \{4\text{-}3,3\text{-}1,1\text{-}2,2\text{-}6,6\text{-}5,5\text{-}7,7\text{-}4\} \end{array}$$

Using GNX with the same parameters (and assuming symmetry so that, for example, 1-2 and 2-1 are the same), the first two offspring steps are as follows:

$$\{1\text{-}6,6\text{-}3,-,-,-,-,-\} \quad \text{and} \quad \{1\text{-}6,6\text{-}3,-,-,-,5\text{-}7,7\text{-}4\}\,.$$

At the third step, only the edge 3-4 in P1 is valid—clearly the edges 4-7 and 7-5 cannot be used again as they would create loops. In fact the only legal alleles left in either parent are 5-2 and 2-1, so the final steps produce

$$\{1\text{-}6,6\text{-}3,3\text{-}4,-,-,5\text{-}7,7\text{-}4\} \quad \text{then} \quad \{1\text{-}6,6\text{-}3,3\text{-}4,5\text{-}2,2\text{-}1,5\text{-}7,7\text{-}4\}\,.$$

In permutation representation this gives

$$1\,6\,3\,4\,7\,5\,2\,,$$

2.3. THE ELEMENTS OF A GENETIC ALGORITHM

which is different from that produced using GNX with the direct permutation representation. (In fact, it is also a clone of the first parent, which may or may not be a good thing!)

Thus, GNX has the merit of constructing a general framework for recombination rather than relying on completely *ad hoc* ideas. A second offspring can be generated by reversing the parental order, and GNX can also easily be generalized further to the case of more than 2 parents. Readers familiar with Glover's idea of *scatter search* [92] will also find some resonance here.

Inversion

Holland's original presentation [124] included an operator that he called *inversion*. The purpose of inversion was not to introduce new material, as with mutation, or recombine different genetic material, as in crossover. Rather it was a re-ordering, applied to an individual string, with the intent of altering the proximity of different genes prior to crossover. In order to make this possible, the locus of each gene must be explicitly encoded along with the allele at that locus. This means that there are many ways of encoding a particular string. For example, the following two strings are the same, where the first row denotes the locus and the second the allele:

$$1\ 2\ 3\ 4\ 5\ 6\ 7 \qquad 2\ 6\ 7\ 4\ 1\ 5\ 3$$
$$1\ 0\ 0\ 1\ 0\ 0\ 1 \qquad 0\ 0\ 1\ 1\ 1\ 0\ 0$$

However, when crossover (1X) is applied to the second string, the group of 1s in the middle is less likely to be broken up than they are in the first. There is an obvious difficulty, however: what happens if the order of the two chromosomes selected for crossover is different? To deal with this, the chromosomes have to be aligned so that the order is the same in each case. This idea and its purpose is linked strongly to Holland's idea of a *schema*, as will be seen in the next chapter.

Implementation issues

Of course, there are also many other practical considerations that influence the implementation of crossover. How often do we apply it? Some practitioners always apply crossover, some never do, while others use a stochastic approach, applying crossover with a probability $\chi < 1$. Do we generate one offspring or two? In many cases natural offspring 'twins' will be created, although in more sophisticated problems it may be that only one offspring arises. If we choose only one from two, how do we do it? In accordance with the stochastic nature of the GA, we may well decide to choose either of the offspring at random. Alternatively, we could bias the decision by making use of some other property such as the relative fitness of the new individuals, or the loss (or gain) in diversity that results from choosing one rather than the other.

2.3.6 Mutation

First we note that in the case when crossover-OR-mutation is used, we must first decide whether any mutation is carried out at all. Assuming that it is, the concept of mutation is even simpler than crossover: a gene (or subset of genes) is chosen randomly and the allele value of the chosen genes is changed. Again, this can easily be represented as a bit-string, so we generate a mask such as

$$0100010$$

using a Bernoulli distribution—i.e., at each locus there is a small probability μ of a mutation occurring (indicated in the mask by a 1). The above example would thus imply that the 2nd and 6th genes are assigned new allele values. However, it appears that there are variant ways of implementing this simple idea that can make a substantial difference to the performance of a GA. The commonest idea seems to be to draw a random number for *every* gene in the string, but this is clearly potentially expensive in terms of computation if the strings are long and the population is large. An efficient alternative is to draw a random variate from a Poisson distribution with parameter λ, where λ is the average number of mutations per chromosome. A common value for λ is 1—in other words, if ℓ is the string length, the mutation rate per gene is $\mu = 1/\ell$. Having decided that there are (say) m mutations, we draw m random numbers (without replacement) uniformly distributed between 1 and ℓ in order to specify the loci where mutation is to take place.

In the case of binary strings, mutation simply means complementing the chosen bit(s). For example, the string

$$1101101$$

with mutation applied at genes 2 and 6, becomes

$$1001111$$

Mutation needs more careful consideration in the case of a q-ary coding. Since there are several possible allele values for each gene, mutation is no longer a simple matter; if we decide to change a particular allele, we must provide some means of deciding what its new value should be. This could be a stochastic choice, but if there is some ordinal relation between allele values, it may be more sensible to restrict the choice to alleles that are close to the current value, or at least to bias the probability distribution in their favour.

It is often suggested that mutation has a somewhat secondary function, that of helping to preserve a reasonable level of population diversity—an insurance policy which enables the process to escape from sub-optimal regions of the solution space, but not all authors agree. The balance between crossover and mutation is often a problem-specific one, and definite guidelines are hard to give.

Several authors have suggested some type of adaptive mutation: for example, Fogarty [77] experimented with different mutation rates at different loci. Reeves [216] varied the mutation probability according to the diversity in the population (measured in terms of the coefficient of variation of fitnesses). More sophisticated procedures are possible, and anecdotal evidence suggests that many authors use varying mutation rates as a diversity maintenance policy.

2.3.7 New population

Holland's original GA assumed a *generational* replacement strategy: selection, recombination and mutation were applied to a population of N chromosomes until a new set of N individuals had been generated. This set then became the new population. From an optimization viewpoint this seems an odd thing to do—we may have spent considerable effort obtaining a good solution, only to run the risk of throwing it away and thus preventing it from taking part in further reproduction. For this reason, De Jong [55] introduced the concepts of *élitism* and *population overlaps*. The ideas are simple—an elitist strategy ensures the survival of the best individual so far by preserving it and replacing only the remaining $(N-1)$ members of the population with new strings. Overlapping populations take this a stage further by replacing only a fraction G (the *generation gap*) of the population at each generation. Finally, taking this to its logical conclusion produces the so-called steady-state or incremental strategies, in which only one new chromosome (or sometimes a pair) is generated at each stage. Davis [47] gives a good general introduction to this type of GA.

Slightly different strategies are commonly used in the ES community, which traditionally designates them either (λ, μ) or $(\lambda + \mu)$. (Note that these λ and μ have a different meaning from those associated with mutation in the previous section.) In the first case, $\mu \geq \lambda$ offspring are generated from λ parents, and the best λ offspring are chosen to start the next generation[4]. For the + strategy, μ offspring are generated and the best λ individuals are chosen from the combined set of parents and offspring.

In the case of incremental reproduction it is also necessary to select members of the population for deletion. Some GAs have assumed that parents are replaced by their children. Many implementations, such as Whitley's GENITOR [301], use the tactic of deleting the worst member(s) of the population, although (as Goldberg and Deb [100] have pointed out) this exerts a very strong selective pressure on the search, which may need fairly large populations and high mutation rates to prevent a rapid loss of diversity. A milder prescription is to select from the worst $P\%$ of the population (for example, Reeves [216] used $P = 50$, i.e., selection from those worse than the median). This is easily implemented when rank-based selection is used. Yet another approach is to base deletion on the *age* of the strings.

Crowding and niching

Another aspect of the generation of a new population is the place of diversity maintenance. This has been a second-order consideration in many of the choices discussed above. It is also possible to make this a first-order question, as with the concepts of crowding and niching. The terms again come from a biological analogy: a *niche* in the natural environment is a set of conditions to which a specific group of phenotypes (a *species*, we might say) is particularly well adapted. In the GA, a niche is treated as a subset of chromosomes that are in some sense similar.

The idea here is that a newly generated chromosome should replace one in its own niche, rather than potentially any one in the population. (This clearly relates only to overlapping populations.) De Jong [55] developed a 'crowding factor' for use in such cases. This was an integer defining the size of a (randomly chosen) subset of the existing population, with which the newly generated chromosome was compared. The one to which it was closest was the one to be deleted.

Other forms of niche generation have been proposed, of which the most well-known is *sharing*. This was first proposed by Goldberg and Richardson [95] who defined a *sharing function* over the population which is then used to modify the fitness of each chromosome. The function could take many

---

[4]The λ, μ strategy is also equivalent to truncation selection as defined earlier with $P = \lambda/100\mu$ [11].

2.3. THE ELEMENTS OF A GENETIC ALGORITHM

forms, but a simple linear function

$$h(d_{ij}) = 1 - d_{ij}/D, \quad d_{ij} < D$$
$$h(d_{ij}) = 0, \quad d_{ij} \geq D$$

was effective. Here d_{ij} is the *distance* between chromosomes i and j, and D is a parameter. The idea is that the sharing function is evaluated for each pair of chromosomes in the population, and then the sum

$$\varsigma_j = \sum_{i \neq j} h(d_{ij})$$

is computed for each chromosome j. Finally the fitness of chromosome j is adjusted by dividing the raw fitness values by ς_j; the adjusted fitness values are then used in place of the original values. The purpose of this is that chromosomes that occur in clusters should have their fitness devalued relative to those which are fairly isolated. The only question is how to measure distance. The obvious measure on the chromosomes is Hamming distance (especially for binary strings). As Hamming distance is so frequently used, we define it formally here:

Definition 2.2 *The* Hamming distance *between two vectors* $\boldsymbol{x}$ *and* $\boldsymbol{y}$ *is given by*

$$d_H(\boldsymbol{x}, \boldsymbol{y}) = \sum_{i=1}^{\ell} [x_i \neq y_i].$$

(This is also a good point to explain the square bracket notation [*expr*], originally introduced by Kenneth Iverson. When square brackets enclose a logical expression *expr*, they denote an *indicator* function, which takes the value 1 if *expr* is true and 0 otherwise. We shall make extensive use of this notation.)

Hamming distance is particularly useful in the case of binary strings, although (as the above definition makes clear) it is not restricted to this case.[5] In the context of niching, however, and as alluded to above in Section 2.2.4, integers that are actually close can be mapped to binary representations that are far apart in Hamming space and *vice-versa*. While this is not a problem with all representations, it may be preferable to define distances in terms of the phenotype rather than the genotype.

---

[5]Hamming distance is frequently used in the analysis of RNA sequences, for example, where the underlying alphabet has cardinality 4.

Many other variations on the theme of niching have been proposed; a recent reference that provides comprehensive details is Horn and Goldberg [131].

2.3.8 Diversity maintenance

No less than John Holland himself has said that 'the very essence of good GA design is retention of diversity' [127]. The effect of selection is to reduce diversity, and some methods can reduce diversity very rapidly. This can be mitigated by having larger populations, or by having greater mutation rates, or by niching or crowding, all of which have some 'natural' basis for the way they are performed. Later, in Section 2.6.2 we shall describe Eshelman's idea of 'incest prevention'. However, we need not be restricted to looking for phenomena that we see occurring in nature.

A popular approach, commonly linked with steady-state or incremental GAs, is to use a 'no duplicates' policy [47]. This means that the offspring are not allowed into the population if they are merely clones of existing individuals. The downside, of course, is the need to compare each current individual with the new candidate, which adds to the computational effort needed—an important consideration with large populations. (In principle, some sort of 'hashing' approach could be used to speed this process up, but whether this was used in [47] is not clear.)

We could also take steps to reduce the chance of cloning before offspring are generated. For instance 1X on the two binary strings

$$1101001$$
$$1100010$$

cannot help but generate clones if the crossover point is any of the first three positions. Booker [26] suggested that before applying crossover, we should examine the selected parents to find suitable crosspoints—the 'reduced surrogate'. A simple way of doing this entails computing an 'exclusive-OR' (XOR) between the parents, so that only positions between the outermost 1s of the XOR string should be considered as crossover points. In the example above, the XOR string is

$$0001011$$

so that, as previously stated, only the last 3 crossover points will give rise to a different string. More generally, whenever there is one or more 0s between a pair of 1s, the same offspring will result from the related crosspoints. In the above example, choosing the 4th or 5th crosspoint gives the same offspring.

2.4 Illustrative Example

It is high time to provide a worked example of the implementation of a GA. As GAs are stochastic in nature, the first thing we need is a random number source. Most computer systems have built-in rand() functions, and that is the usual method of generating random numbers. The table below provides some random numbers that we shall use in what follows.

Table 2.1: Some uniformly distributed random numbers.

0.714	0.268	0.284	0.770	0.575	...				
0.185	0.367	0.798	0.236	0.174	0.562	0.385	0.096	0.441	...

Suppose we wish to find the optimum of the cubic function defined in Example 2.4 above. Let us assume we have an initial population of size 5. If we generate a random number r, then we assign each allele the value 0 if $r < 0.5$ and a 1 otherwise. Thus the first row in Table 2.1 generates the genotype string (1 0 0 1 1). We can decode this to get the phenotype $x = 16 + 2 + 1 = 19$, and evaluate $f(19) = 2399$. Following the same procedure with 4 more sets of random numbers produces the first generation: the initial population shown in Table 2.2. (Note that the random numbers involved have been omitted).

The final column in Table 2.2 is the fitness-proportional selection probability, from which we can derive the following cumulative distribution for roulette-wheel selection:

Rand. no.	0.206	0.483	0.527	0.682	1.000
String	1	2	3	4	5

Thus, returning to Table 2.1, our next random number is 0.185, which implies that we select string 1; the next is 0.367, which (as it lies between 0.206 and 0.483) means that we select string 2. Now we perform crossover and mutation on these strings. To apply one-point crossover, we have to choose a crosspoint; this is selected with equal probability from the numbers 1, 2, 3, 4, using the following distribution:

Rand. no.	0.25	0.50	0.75	1.000
Crosspoint	1	2	3	4

As the next random number is 0.798, the chosen crosspoint is 4. If we cross the strings (1 0 0 1 1) and (0 0 1 0 1) at the 4th crosspoint, the resulting

Table 2.2: Initial population and associated calculations.

No.	String	x	$f(x)$	$\Pr[select]$
1	1 0 0 1 1	19	2399	0.206
2	0 0 1 0 1	5	3225	0.277
3	1 1 0 1 0	26	516	0.044
4	1 0 1 0 1	21	1801	0.155
5	0 1 1 1 0	14	3684	0.317
Average fitness			2325	

strings are (1 0 0 1 1) and (0 0 1 0 1) again! This illustrates one of the problems that we mentioned above—crossover can produce offspring that are merely clones of their parents. (Section 2.3.8 discussed some consequences of this, and how to prevent it.) For the moment, we shall choose one of these strings with equal probability: if the next random number is less than 0.5, we choose the first, otherwise the second. As our next random number from Table 2.1 is 0.236, we shall proceed with the first string, i.e. (1 0 0 1 1).

Now we need to apply the mutation operator. We shall suppose that the mutation probability (or rate) μ in this case is 0.10 for each locus of the string. Since the next random number is 0.174 (> 0.1), there is no change to the allele value at locus 1. Similarly, the alleles at loci 2 and 3 remain the same, but at locus 4, the random number is only 0.096, so the allele value for this gene changes from 1 to 0. There is no change at locus 5, so the final string is (1 0 0 0 1), which decodes to $x = 17$ with fitness $f(17) = 2973$.

This procedure can then be repeated to produce a second generation, as seen in Table 2.3.

The search would continue by re-calculating the selection probabilities in order to carry out a new cycle of selection, crossover and mutation. This is a very simple GA, but it has been effective in raising the population average quite substantially, and it has actually found the optimum at step 2.

Table 2.3: Second generation population.

Step	Parent 1	Parent 2	Crossover point	Mutation?	Offspring String	$f(x)$
1	1	2	4	NNNYN	1 0 0 0 1	2973
2	5	3	2	NNNNN	0 1 0 1 0	4100
3	5	2	3	NNNNN	0 1 1 0 1	3857
4	4	2	1	NYNNN	1 1 1 0 1	129
5	2	5	4	NNNNN	0 0 1 0 0	2804
Average fitness						2773

2.5 Practical Considerations

2.5.1 The importance of being random

GAs are highly stochastic algorithms, needing many random numbers, as can be seen from the above example, and this has some important implications. One aspect of this has already been mentioned several times—the distortion that can arise from using pure random sampling with replacement, and some methods of variance reduction as used in the simulation literature have been mentioned. Another important aspect, but one not widely publicised, is the need to use a good pseudo-random number generator (PRNG). Ross [240] posted a message to the GA List Digest that contained a salutary lesson to those who rely on the PRNGs provided with standard software. However, Meysenberg and Foster [165] conducted some extensive tests that failed to show any systematic effect on GA performance due to different PRNGs, so the case cited by Ross may be an isolated instance. Nonetheless, it is best to be prepared, and the best advice is to use a PRNG that has been thoroughly tested, such as those described in the *Numerical Recipes* books [191].

2.5.2 Constrained optimization

So far we have ignored the question of constraints on the decision variables of our problem, or implicitly assumed that the only constraints that matter are 'box' constraints on individual variables, which we may make use of in choosing our representation, as in Section 2.2.4. However, in many cases there may be quite complicated sets of constraints on the permissible values of the phenotypic variables. The essence of the difficulty is clear: there is rarely a guarantee (at least with simple operators and codings) that two feasible parent solutions will provide feasible offspring. In some problems it may even be difficult to find a reasonably diverse initial population of feasible solutions.

A popular strategy is to modify the fitness function by adding penalty terms. Goldberg [96] gives the impression that this is a simple matter, and it is often worth trying it, but more extensive experience has suggested that this is seldom the best approach. We briefly discuss some of the alternatives below, but a far more detailed consideration of this question may be found in the paper by Michalewicz and Schoenauer [167].

Penalties

The most obvious solution to the problem of constraints is simply to ignore them. If an infeasible solution is encountered, it is not allowed to enter the population (i.e., a 'death penalty'). Unfortunately, this fails to recognize that the degree of infeasibility does supply some information. It is common to find the global optimum on or near a constraint boundary, so that solutions that are slightly infeasible may actually be fairly close to the optimum. (This is the argument made, for example, by proponents of the concept of *strategic oscillation* in tabu search [91].)

Thus, the obvious next step is to modify the objective function by incorporating penalties. There are cases in which this works well, where theoretical considerations permit an appropriate choice of penalty, but a naive attempt to use penalties often fails. If the penalty is too small, many infeasible solutions are allowed to propagate; if it is too large, the search is confined to the interior of the search space, far from the boundaries of the feasible region. Using variable penalties can be helpful [167].

Repair

Another popular solution to GAs for constrained problems is to accept an infeasible solution, but to repair it before inserting it in the population. Some biological justification is often claimed for this approach, along the lines that the infeasible solution should be treated as an immature individual that, although initially demonstrably unfit for its purpose, can nevertheless develop over time into a perfectly reasonable solution. This analogy has also been applied to methods that use the GA offspring as the springboard for a local search. (Such hybrids have been given a more formal specification

2.5. PRACTICAL CONSIDERATIONS 53

under the name of memetic algorithms by Radcliffe and Surry [203].)

In such cases, there is also the practical problem of deciding whether it should be the original offspring or the repaired version that should be used in the next generation. Inserting the repaired offspring means curtailing the exploration of a promising part of the search space. Davis and colleagues [48, 184] report some interesting experimental evidence, which suggests that the decision as to which to insert should be a stochastic one, with a fairly small probability of 5-10% of inserting the repaired version. There is also a good case for a mixture of approaches, where the original infeasible solution enters the population, but it is given the repaired version's fitness. More will be said about this idea below in Section 2.5.3.

Multiple objectives

Many authors have suggested using infeasibility as a secondary objective, thus avoiding the introduction of penalties. Recent work by Chu and Beasley [38] provides a good example of this idea in the context of set partitioning. In addition to the usual fitness measure (based on the true objective function), they define a secondary measure that they call *unfitness* to represent the degree of infeasibility present in a given solution. The selection scheme for reproduction favoured strings that were maximally compatible, while selection for deletion was carried out by trying to find an existing population member that was dominated by the new offspring—in the sense either that its fitness is greater while its unfitness is no worse, or that its fitness is no worse while its unfitness is less. Multiple objective GAs (MOGAs) and their application have themselves been thoroughly reviewed by Fonseca and Fleming [84].

Modified operators

One approach to alleviating the difficulties encountered by the standard operators is to design operators that are appropriate for the problem. The edge-recombination operator [304] for the TSP (Section 2.3.5) is a typical example.

Modified formulation

Another approach that has proved fruitful in some cases is to modify the problem formulation. Reeves [219] found good solutions to bin-packing problems by reformulating the problem. Rather than using the subset-selection interpretation, bin packing was treated as a sequencing problem, with the sequence being decoded by using an on-line heuristic.

This type of problem re-formulation has also been used in other applications. Schaffer and Eshelman [255] describe a similar approach to an extremely complicated line-balancing problem in the production of printed circuit boards. The GA was used to find a good sequence to feed to a set of heuristics that produced the final solution.

2.5.3 Hybridization

The effectiveness of a GA can nearly always be enhanced by hybridization with other heuristics. While one of the main benefits of using a GA is often professed to be its domain independence, this is also a common reason for GAs performing ineffectively. Introducing problem-specific knowledge can improve solution quality considerably.

Hybridization can be carried out in various ways. The idea of seeding a population (as discussed above) with the results of applying another problem-specific heuristic can be viewed as a type of hybridization, but the most common approach is to integrate other heuristics such as neighbourhood search or simulated annealing into the normal GA cycle. Once again there are several different options, but a typical idea is to apply some standard mutation and crossover operators to generate offspring, which are then allowed to undergo further modification using NS before being inserted into the next generation. How many of them undergo such improvement, whether the NS phase is carried on to a local optimum, whether steepest ascent or something speedier is used, how population diversity is maintained in the fact of fierce convergence pressure—these are examples of questions that have to be addressed in any particular implementation.

Lamarck or Baldwin?

Those who have some knowledge of the history of evolutionary theory will see an obvious association of a typical hybridization scheme with Lamarck's idea of the inheritance of acquired characteristics, but there is another approach, associated with Baldwin [16]. An important early paper connecting these ideas to GAs is that by Hinton and Nowlan [120]. In such cases, the modification is allowed to proceed in accordance with whatever rules have been decided. However, it is not the 'mature' individual that is inserted in the next generation, but the immature or original offspring from which it was obtained. The essential characteristic of the 'Baldwin effect' is that this immature individual then takes the fitness of its mature counterpart [308]. In a sense, the genotype is allowed to receive the credit for what the phenotype has 'learned' from its environment, so that genotypes that possess sufficient flexibility to permit learning will benefit relative to those that do not. Several comparisons have suggested that this is more effective than the Lamarckian approach; in particular, a complete issue of the journal *Evolutionary Computation* was devoted to this idea for the anniversary of Baldwin's original paper [281].

2.6 Special Cases

While this chapter has covered the basic principles of GAs, it should be very clear by now that there are many ways of implementing a GA, and it is likely that—apart from those people who simply use one of the many publicly-available codes, such as Goldberg's simple GA [96]—everybody's GA is unique! There are a few GAs that have a certain importance, either in the primacy of their ideas (such as GENITOR [301], which first systematically advocated ranking), or in the rather particular way in which the GA is implemented. Three of these will be discussed in a little more detail, in order to give a flavour of what is possible.

2.6.1 Breeder GA

Mühlenbein and Schlierkamp-Voosen [175] based their GA on the analogy we mentioned in Chapter 1—to artificial selection as performed by human breeders of plants and animals. A well tried and tested mixture of theory and statistical experiments had been developed over the course of the last century, dealing with concepts of selection intensity and heritability. On this foundation, the *Breeder Genetic Algorithm* (BGA) was constructed. It is similar in spirit to the classical GA of Holland, but is applicable to continuous problems directly without the need for a discrete genotype. (Of course, the BGA can also be used for naturally discrete problems.) Mühlenbein and Schlierkamp-Voosen defined the BGA by means of 5 inter-related parameters:

- String length

- Initial frequency of desirable allele

- Population size

- Mutation rate

- Selection intensity

The first two of these parameters are fixed by the representation, and by the initial population construction respectively. The other three are the ones that really control the process. The notion of selection intensity has been described above, and in this case, it relates to truncation selection[6]. The usual assumption is that phenotypic fitness is normally distributed, and from this I can be determined for different values of the truncation threshold P.

Mutation in the BGA is carried out at a rate of $1/$(string length)—a rate that has often been claimed[7] to be 'optimal': the sense in which it may be considered optimal is discussed in [175], where it is shown that there will also be an optimal population size for a given selection intensity and mutation rate. The fundamental idea here is that the requirements of recombination and mutation are quite different. Mutation is more efficient in relatively small populations, while for recombination to work effectively the population must be large enough to possess sufficient numbers of advantageous alleles to try out in different combinations. These observations are not surprising in themselves, but the quantitative analysis in [175] shows useful insights. In particular, these results appear to show that mutation has a rather more influential rôle than it is sometimes accorded.

There is a *caveat*: the derivation of these results depend heavily on the assumption that the fitness function is additive (and in many cases, are actually only derived for the *Onemax* function[8])—i.e., precisely those functions for which a GA is unlikely to be the best method to use! However, they are assumed to be valid guidelines for other more complicated functions, and for alternative forms of crossover and selection, and the experimental evidence presented in [175] would appear to support this conclusion.

---

[6]A slightly different form of truncation selection is also sometimes used in breeding, where the best string takes place in all matings—an idea that Fairley [75] called Herd Leader breeding. This has recently been 're-discovered' as the 'stud' GA [147]. Others (perhaps more politically correct) have preferred to call it 'queen bee' selection!

[7]Bremermann [30] was apparently the first to derive this result.

[8]The *Onemax* function is defined in Appendix A.

2.6.2 CHC

Eshelman's CHC[9] algorithm [73] takes a quite different approach from the BGA. Here recombination is heavily emphasized, while mutation is used purely as a means to restart the algorithm when crossover's progress has more or less ceased. The key principle for this version of a GA is the maintenance of diversity. This is fostered by the use of a highly disruptive version of UX (which Eshelman calls HUX), and by explicitly preventing 'incest'—the mating of strings whose similarity is above some specified threshold. To allow for the fact that populations cannot help but get more similar under the influence of selection, this threshold is allowed to change as the number of generations increases.

Selection is accomplished by what Eshelman calls a population élitist strategy, which is essentially the $\lambda + \mu$ strategy used by proponents of evolution strategies, with the difference that (unlike ES) the number of offspring μ generated in CHC is variable. All members of the current population are paired randomly, but not all are mated because of the requirement for incest prevention. Only the offspring of these matings are allowed to compete with the current population for survival. Thus, following each selection/recombination step, the best λ of the $\lambda + \mu (\leq 2\lambda)$ strings are chosen to form the next generation.

HUX forces the exchange of exactly half of the non-matching bits (or nearly half, if this number is not an integer), so that the offspring are equidistant from their parents. To those who might object on the grounds of violating the 'building block hypothesis' (which we shall cover in some depth in Chapter 3), Eshelman counters that such violations are relatively insignificant, and that the radical nature of the search accomplished by CHC is a worthwhile gain to put against this hypothetical loss.

Finally, without mutation, CHC is bound to converge to a point where the search stagnates unless the ban on incest is lifted completely. Eshelman takes the view that this is the point at which mutation should be employed. The mutation rate (or *divergence rate* as Eshelman calls it) recommended would seem extraordinarily high (e.g., 0.35 per gene) in the context of ordinary GAs, but of course the purpose here is different. Furthermore, mutation is applied to one string only—the best one, and it is applied $(\lambda - 1)$ times. The new population thus consists of a single copy of the best string, together with the $(\lambda - 1)$ mutated offspring.

With this new population a new recombination phase begins, with the

---

[9]According to Whitley [307], CHC stands for Cross-generational selection, Heterogeneous recombination and Cataclysmic mutation.

incest threshold reset to a value that depends on the mutation rate. In most of the experiments on benchmark problem instances reported in [73], CHC was run until a global optimum was found, but in general the process has no obvious stopping criterion. Thus some termination condition (e.g., a maximum number of restarts, or a limit on the number of function evaluations) would have to be imposed externally. The results of the experiments certainly demonstrated that CHC with a fixed population size of only 50 and a fixed divergence rate of 0.35 tends to outperform a traditional GA, even when that GA's parameter set is 'optimal', across a range of different types of problem. Pseudocode for CHC is given in [73], and it is relatively easy to program it using this information.

2.6.3 GAVaPS

Population size has often been identified as a critical parameter in determining whether or not a GA will work efficiently (or even at all). Without sufficient diversity, the search will get nowhere fast, but too large a population will slow it to a crawl and the available computational effort will be exhausted. CHC seems to provide a counter example, but only inasmuch as it approaches diversity maintenance in a different way. For standard GAs, the question of population size remains.

One interesting approach is that of Michalewicz, who allows the size of the population to vary by focusing on the *age* of a string. Age has been mentioned before as a possible criterion for deletion in steady-state GAs, but in GAVaPS [166] the *lifetime* of a string is incorporated as an explicit parameter. Once the age of a string exceeds its lifetime, it is eliminated from the population. At each 'generation' a fixed proportion of the population is selected for recombination and crossover in the usual way, and the lifetime of each offspring is set. This value is determined (within externally imposed limits) by reference to population statistics: principally, the mean, minimum and maximum fitness of the current population; in some versions, the best and worst fitnesses seen so far are also used.

Experimental evidence presented in [166] suggests that this approach, with minimum and maximum lifetimes set at 1 and 7 respectively, outperforms a traditional GA with a fixed population. Actual population sizes show an interesting profile: early in the search the population becomes comparatively large, but later it shrinks to a much smaller value. Similar behaviour is also seen on a smaller scale whenever new high-fitness strings are discovered. It is argued that this shows an efficient allocation of effort, allowing sufficient exploration to be balanced by effective exploitation. Of

2.7. SUMMARY

course, there is a potential drawback, in that at least two new parameters (minimum and maximum lifetimes) have to be specified, as well as the existence of further scope for uncertainty as to what form the lifetime function should take.

2.7 Summary

This chapter has discussed the basic steps involved in implementing a genetic algorithm, and have outlined some special cases. It should by now be clear that GAs provide an unrivalled opportunity for variation in implementation, each element having so many possible strategies, which in turn rely on so many quantitative parameters whose values can influence the GA's performance quite markedly. Three very different individual GAs have been described to reinforce this point.

While this wealth of potential is part of the appeal of the GA paradigm, it is also somewhat frustrating, both for the practitioner and for the theorist. We hesitate to make a firm recommendation (someone will only disagree!), but would tentatively suggest that

- An initial population of about 50 should contain sufficient alleles for the GA to make useful progress.

- Prior knowledge should be used along with randomization in choosing the initial strings.

- SUS and tournament selection are more efficient than the basic RWS.

- Incremental reproduction and replacement usually make better use of resources than the generational approach.

- Two-point or uniform crossover have less positional bias than 1X.

- An adaptive mutation rate appropriate to the application should be used, but if in doubt, a fixed rate of $1/\ell$ is a reasonable choice.

- Diversity maintenance should be prominent in any implementation.

- Hybridization should be used wherever possible; a GA is not a black box, but should use any problem-specific information available.

- GAs are stochastic; there should be several replicate runs in any application.

From the practical side, some steps have been proposed to reduce the complexities of implementation. In the ES community, for instance, it has become commonplace to make use of self-adaptive mutation rates (see [12, 21] for a review of some of these developments). Recently Bäck et al. [14] have reported some experiments with a 'parameter-less' GA. Actually, as in the case of GAVaPS, with which it shares some similarities, parameters have not been eliminated—rather, they are replaced with different ones. However, it is hoped that the resulting algorithm might be less sensitive than traditional GAs.

From the theoretical side, it would appear that probing more deeply into questions of how and why GAs work is vastly complicated by the existence of such a diversity of approaches. However, the basic ideas of population-based selection, recombination and mutation are common to everything that can be called a GA. In the second part of this book we shall consider some of the theoretical perspectives that have been applied, and we shall find that mostly they deal with general principles of search strategy using recombination and mutation without being overly concerned with the details. We begin in the next chapter by considering Holland's original arguments, which make use of the key ideas of schemata.

2.8 Bibliographic Notes

Again, John Holland's seminal work [124, 126] is well worth reading for an insight into the development of the basic notions of a GA. David Goldberg's book [96] is deservedly well known for its clarity and careful organisation. Darrell Whitley's tutorial paper [307] is an excellent summary of the basic principles of a GA, while other worthwhile sources of information are the books by Thomas Bäck [12] and Melanie Mitchell [170].

More idiosyncratic but nonetheless useful and interesting works are those by Michalewicz [166] and Falkenauer [76]. Ken De Jong's thesis [55], which did so much to stimulate later enhancements of GAs is unfortunately unpublished[10], but he has recently produced a book [60] on evolutionary computation.

Genetic algorithms have developed as a subject of research in large measure because of a few series of major conferences. At first, these were biennial, with two series alternating between the USA and Europe. As the field developed, other conferences began, and there are now many confer-

---

[10]However, scanned PDF files can be downloaded from George Mason University—see the URL given in the bibliography.

2.8. BIBLIOGRAPHIC NOTES

ences each year in most parts of the world. The details of the proceedings of most of these conferences are listed in the bibliography (items [321]-[350]); readers interested in theoretical aspects of GAs will find that the biennial workshops under the title *Foundations of Genetic Algorithms* (items [345]-[350]) contain much to supplement the relevant subject matter of the next 6 chapters of this book.

Exercises

1. A real-valued problem is to be solved by using a binary coding of the variables. There are 6 variables; x_1 is known to lie in the interval $[0, 100]$, the rest are thought to be somewhere between -1000 and 1000. The required precision is 10^{-2} in the cases of x_1, x_4 and x_5, and 10^{-1} for the rest. Construct a suitable genotype-phenotype mapping. How long is the binary string? What is the binary representation of the point $(20, -570, 120, -33, 33, 33)$?

2. Prove the result
$$N \approx 1 + \log\left(-\ell/\ln P_2^*\right)/\log 2.$$
for the minimal population size, where the parameters have the meaning defined in Section 2.3.1.

3. Determine the values a, b necessary for linear scaling as defined in Section 2.3.4 with selection pressure ϕ.

4. Prove the formulae stated in Section 2.3.4 for linear ranking:
$$\beta = \frac{2(\phi - 1)}{N(N-1)} \quad \text{and} \quad \alpha = \frac{2N - \phi(N+1)}{N(N-1)}$$

5. (continuation) Verify that
$$k = \frac{-(2\alpha + \beta) + \sqrt{(2\alpha + \beta)^2 + 4\beta r}}{2\beta}$$
where r is a $U(0, 1)$ random number.

6. If strings are ranked in ascending order of fitness, show that the probability of selecting the kth ranked string in soft tournament selection with $\tau = 2$ is
$$p\frac{2(k-1)}{N(N-1)} + (1-p)\frac{2(N-k)}{N(N-1)}$$
where p is the probability of choosing the better string in the tournament, and hence prove its equivalence to linear ranking.

7. Geometric ranking (sometimes called exponential ranking) assumes that the probability of selecting the kth ranked individual (in ascending fitness order) is
$$\mathbf{Pr}[k] = \alpha\beta^k$$

2.8. BIBLIOGRAPHIC NOTES

Deduce suitable values of α and β assuming a selection pressure ϕ, and evaluate them for the case $N = 5, \phi = 2$. Compare the probabilities to those obtained for linear ranking with the same parameters.

8. Show that in tournament selection with $\tau > 2$ the probability of selecting the kth ranked string is

$$\mathbf{Pr}[k] = \left(\frac{k}{N}\right)^\tau - \left(\frac{k-1}{N}\right)^\tau.$$

By comparing the probability $\mathbf{Pr}[N]$ with that found in Exercise 7, deduce that this is approximately the same as geometric ranking if

$$\beta = \left(\frac{N}{N-1}\right)^\tau.$$

9. Generate and compare examples of uniform (binary mask) order crossover, edge-recombination and GNX in both vertex and edge form for the following TSP permutations:

    P1   1 8 2 7 4 5 3 6 9   P2   4 6 1 9 7 3 8 5 2

Chapter 3

Schema Theory

Holland's original insights into the analysis of GA behaviour have come in for some criticism over the years, but some of the basic ideas have proved fruitful, and it is entirely appropriate to review them at this point. As we have already hinted, both he and Goldberg have argued strongly for the use of binary encoding: in the terminology of Chapter 2—for the use of the search space $\mathcal{A}^\ell$ (or a subset $\mathcal{X}$ of it), rather than $\mathcal{V}$ itself. His explanation for this hinges on 3 main ideas, which we shall discuss at length in this chapter.

3.1 Schemata

However, prior to such a discussion we need to introduce his central concept of a *schema*[1].

Definition 3.1 *A schema is a subset of the space $\mathcal{A}^\ell$ in which all the strings share a particular set of defined values.*

This can be represented by using the alphabet $\mathcal{A} \cup *$, where the $*$ symbol is a 'wild card', i.e. it can be replaced by any of the possible alleles. In the binary case $(1**1)$, for example, represents the subset of the 4-dimensional hypercube $\{0,1\}^4$ in which both the first and last genes take the value 1, i.e., the strings $\{(1\,0\,0\,1), (1\,0\,1\,1), (1\,1\,0\,1), (1\,1\,1\,1)\}$.

Schemata can thus be thought of in set-theoretic terms, as defining subsets of similar chromosomes, or geometrically, as defining *hyperplanes* in

---

[1] The word comes from the past tense of the Greek verb $\epsilon\chi\omega$ (*echo*, to have), whence it came to mean shape or form; its plural is *schemata*.

ℓ-dimensional space. Another interpretation is in terms of periodic 'functions' of different 'frequencies'—for example, as illustrated in Figure 3.1 the schema $(* \cdots * 1)$ represents the odd integers in the case where the chromosome (genotype) encodes integer values (phenotype). As we shall see later, in the case of binary strings, these 'functions' have a particular form that is fundamental to much theoretical work.

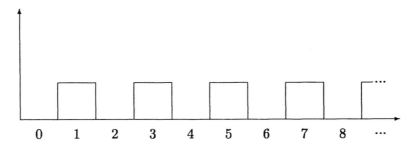

Figure 3.1: Possible integer values of the schema $(* \cdots * 1)$.

It is clear that any specified chromosome is an instance of many schemata. In general, if the string has length ℓ, each chromosome is an instance of $|\mathcal{A}|^\ell$ distinct schemata, since at each locus it can take either a $*$ or its actual value. (We assume that the 'degenerate' cases—the string itself and the all $*$ string—are also schemata.) As a consequence of this, each time we evaluate the fitness of a given chromosome, we are gathering information about the average fitness of each of the schemata of which it is an instance. In the binary case, a population of size N could contain $N2^\ell$ schemata, but of course in practice there will be considerable overlapping between strings so that not all schemata will be equally represented, and some may have no representatives at all. (In fact we want them to have unequal representation, as the GA is suppose to focus attention on the ones that are fitter.)

3.1.1 Intrinsic or implicit parallelism

Holland claimed that in the sense that we are testing a large number of possibilities by means of a few trials, a GA possesses a property that he originally called *intrinsic parallelism*. Some other authors have used the alternative term *implicit* parallelism, and Holland himself later suggested [126] that this be adopted, reserving the term intrinsic to refer to the opportunity for a software *implementation* of GAs in parallel.

The question is of course, how many schemata are being processed? To

3.1. SCHEMATA

answer this in general is a formidable task, but we can simplify by first restricting the question to those schemata which have a probability $1 - \eta$ of surviving.

We also need the notions of the *length* and *order* of a schema. These are respectively the *distance* between the first and last defined (i.e. non-∗) positions on the schema, and the *number* of defined positions. Thus, the schema (1∗ ∗ 1) has length 3 and order 2.

The value η can be thought of as a 'transcription error'; it is of course a function of schema fitness, length and order, as will be seen in the next section where we deal with the the Schema Theorem. For the moment, we will assume that this restriction implies that we need only consider schemata of length $\leq 2m$, say (where m is a function of η), and of order m.

In order to estimate the number of schemata satisfying these conditions, we consider a 'window' of $2m$ contiguous loci counting from the leftmost position. There are $\binom{2m}{m}$ possible choices of defining loci for an order m schema in this window, each of which would give rise to 2^m actual schemata. (For example, (f ∗ f ∗ · · · ∗), where f is a fixed value of 0 or 1, represents 4 schemata corresponding to the 4 ways of choosing the fixed values.) We now successively move this window 1 step to the right, observing that each time there will be $\binom{2m-1}{m-1}$ sets of defining loci that have not been counted in the previous window. Altogether, we can move this window 1 step to the right $(\ell - 2m)$ times, so there are

$$\left[\binom{2m}{m} + (\ell - 2m)\binom{2m-1}{m-1}\right] 2^m = (\ell - 2m + 2)\binom{2m}{m} 2^{m-1}$$

possible schemata per string $\approx 2^{3m-1}/\sqrt{\pi m}$, using Stirling's approximation for the factorials.

Of course, in a given string, all 2^m schemata are not represented for a particular choice of defining loci, so lastly, we need to estimate how many schemata meeting the prescribed conditions will occur in a population of size N. On the assumption that the members of the population are chosen using uniform random sampling, a population of size $N = \nu 2^m$ (for ν a small integer value) would be *expected* to have ν instances of each of these order m schemata, so that with this definition of N, we can estimate the number of schemata of length $2m$ or less, and order m, as $\mathcal{O}(N^3)$. In fact, this is really an under-estimate, since there are many other schemata we have not counted at all, so this establishes a lower bound.

3.2 The Schema Theorem

But how relevant is this idea of implicit parallelism? It is clearly impossible to store the estimated average fitness values of all these schemata explicitly, nor is it obvious how to exploit such information even if we could. The second plank of Holland's argument was to show that by the application of genetic operators such as crossover and mutation, each schema represented in the current population will increase or decrease in line with its relative fitness, independently of what happens to other schemata.

In mathematical terms, he proved what has become known as the Schema Theorem. We shall prove this theorem in the case of fitness-proportional selection. For this case, the *fitness ratio* $r(S,t) = f(S,t)/\bar{f}(t)$ is an important parameter, expressing the average fitness of a schema relative to the average fitness of the population. We have of course assumed here that the fitness of a schema is the average fitness of all instances of that schema in the population $P(t)$, i.e.,

$$f(S,t) = \frac{\sum_{x \in S \cap P(t)} f(x)}{|S \cap P(t)|},$$

while

$$\bar{f}(t) = \sum_{x \in P(t)} f(x)/|P(t)|.$$

(Note that in general $P(t)$ is a *multiset*, i.e., multiple occurrences of individual chromosomes are possible—indeed, highly probable.) We approach this theorem by proving the following results:

Lemma 3.1 *Under a reproductive plan in which a parent is selected in proportion to its fitness, the expected number of instances of a schema S at time $t+1$ is given by*

$$E[N(S,t+1)] = r(S,t)N(S,t)$$

where $N(S,t)$ is the number of instances of S at time t.

Proof The result follows directly from the definition of fitness-proportional selection. □

Of course, this only describes what happens under selection. If crossover and mutation are involved, instances of schema S may become instances of some other schema S' (destruction). At the same time, instances of other

3.2. THE SCHEMA THEOREM

schemata could recombine or mutate to become instances of S (construction), but Holland chose to neglect them, so that he could state a result about S independently of all other schemata.

Lemma 3.2 *If one-point crossover is applied at time t with probability χ to a schema S of length $l(S)$, then the probability that S will be represented in the population at time $t+1$ is bounded below by*

$$1 - \chi \frac{l(S)}{\ell - 1} P_{diff}(S, t)$$

where ℓ is the length of the chromosome, and $P_{diff}(S,t)$ is the probability that the second parent is an instance of a different schema.

Proof The probability that schema S will be destroyed by the crossover operator is easily seen to be the product of the probability that crossover is applied, χ, and the probability that the crossover site lies within the length of the schema, $l(S)/(\ell - 1)$.

If the other parent is itself an instance of S, then S will survive regardless of the crossover site; hence this must be multiplied by $P_{diff}(S,t)$ to give the probability of destruction.

Hence the probability of non-destruction is

$$1 - \chi \frac{l(S)}{\ell - 1} P_{diff}(S, t)$$

Finally, the probability of non-destruction is clearly only a lower bound on the probability of representation, since it is also possible for S to appear at time $t+1$ as a result of crossover of instances of two different schemata S' and S''. □

For most selection and mating schemes, $P_{diff}(S,t)$ will depend only on S, so this is still independent of other members of the population. However, since the maximum value that $P_{diff}(S,t)$ can take is 1, the lower bound could be replaced by

$$1 - \chi \frac{l(S)}{\ell - 1}$$

which removes any possibility that what happens to S is dependent on the rest of the population.

Lemma 3.3 *If mutation is applied at time t with probability μ per gene to a schema S of order $k(S)$, then the probability that S will be represented in the population at time $t+1$ is bounded below by*

$$1 - \mu k(S)$$

Proof From the definition of mutation, the probability that mutation fails to destroy S is given by

$$(1-\mu)^{k(S)} \geq 1 - \mu k(S)$$

□

Combining these results together we obtain the following theorem:

Theorem 3.4 (The Schema Theorem) *Using a reproductive plan as defined above in which the probabilities of crossover and mutation are χ and μ respectively, and schema S of order $k(S)$, length $l(S)$ has fitness ratio $r(S,t)$ at time t, then the expected number of representatives of schema S at time $t+1$ is given by*

$$E[N(S,t+1)] \geq \left\{1 - \chi\frac{l(S)}{\ell-1}P_{diff}(S,t) - \mu k(S)\right\}r(S,t)N(S,t) \quad (3.1)$$

Note 1 As noted above, the $P_{diff}(S,t)$ term can be ignored to leave the Schema Theorem in the form

$$E[N(S,t+1)] \geq \left\{1 - \chi\frac{l(S)}{\ell-1} - \mu k(S)\right\}r(S,t)N(S,t) \quad (3.2)$$

which is a slightly simpler, and perhaps more commonly seen variation.

Note 2 The above statement of the theorem is still different from that sometimes found. In particular, the left-hand-side is explicitly stated in terms of *expectation*, a fact which is often insufficiently stressed; also the effect of mutation is included. Secondly, the dynamic dependence of the fitnesses on the population is emphasized—again, something that not all statements of the Schema Theorem make clear.

Note 3 While this development has assumed a generational GA with gene-wise mutation, one-point crossover and fitness-proportional selection, similar statements can be found for other types of selection, crossover and mutation. A 'generic' version of the Schema Theorem would look like

$$E[N(S,t+1)] \geq \{1 - \eta(S,t)\}\sigma(S,t)N(S,t) \quad (3.3)$$

where $\eta(\cdot)$ represents a 'loss' term relating to the 'transcription error' effects of recombination and mutation, while $\sigma(\cdot)$ is a 'selection coefficient'. This

3.2. THE SCHEMA THEOREM

way of stating the Schema Theorem points up Holland's basic idea that there are two competing forces in a GA—one increases the representation of good schemata in line with their fitness, while the other tends to break them up. So what characterizes the survivors?

3.2.1 Building block hypothesis

Consideration of the expression (3.2) shows that the representation of S in the population is expected to increase provided that

$$r(S,t) \geq 1 + \chi \frac{l(S)}{\ell - 1} + \mu k(S) \tag{3.4}$$

Thus, short low-order schemata will increase their representation in the next population provided their fitness ratio is slightly more than 1; long and/or high-order schemata need much higher ratios.

The ideal situations for a GA would therefore seem to be those where short, low-order schemata or 'building blocks' are able to combine with each other to form better solutions. The assumption that this is the GA's *modus operandi* is what Goldberg [96] calls the *building block hypothesis*. This might be a reasonable assumption in some problems, but how it can be ensured in general is the real question—one we shall consider a little later.

3.2.2 The two-armed bandit analogy

Holland also attempted to model schema processing (or hyperplane competitions) by means of an analogy to a stochastic two-armed bandit problem. This is a well-known statistical problem: we are given two 'levers', which if pulled give 'payoff' values according to different probability distributions. The problem is to use the results of previous pulls in order to maximize the overall future expected payoff. In [124] it is argued that a GA approximates an 'optimal' strategy, which allocates an exponentially increasing number of trials to the observed better lever; this is then used to contend for the supposed efficiency of a GA in distinguishing between competing schemata or hyperplanes.

Of course, this analogy assumes we might be solving problems of the form 'is $(1****)$ better than $(0****)$?'. In practice, we need to solve several of these problems (for example, 'is $(*1***)$ also better than $(*0***)$?'), and unless they are independent, the sequence in which they are solved might be important. Further, we are usually interested in higher-order schemata. (If we are not, then the problem is by default a linear one, in which case we should hardly be using a GA to solve it!) Thus there are really several bandits, which have rather more than two arms. Nevertheless, Holland argued that we can place a lower bound on the loss caused by making the 'wrong' decision between several schemata, so that the theory could still be applied.

3.2.3 Principle of minimal alphabets

Holland also used his schema-based theory to speculate on the 'optimality' of binary encodings. The number of potential schemata is $(|\mathcal{A}|+1)^\ell$, where ℓ is the length of the string used to encode the decision variables, which of course depends inversely on the cardinality of the alphabet $\mathcal{A}$. Clearly the exponent ℓ is more important than the factor $|\mathcal{A}|+1$, and in fact, the maximum number of schemata will arise when ℓ is as large as possible, which occurs when the alphabet has as few symbols as possible, i.e., two. This has become known as the 'principle of minimal alphabets' (PMA).

3.2.4 Does it work? Empirical evidence

We shall use the example of Section 2.4 again to illustrate what the underlying schema averages may have to tell us. Table 3.1 contains the average values for all the order-1 schemata given by the strings in the populations recorded in Tables 2.2 and 2.3.

This table reveals some interesting points for discussion. Firstly, the 'predictions' of the order-1 schema averages (the lowest order building blocks) in the initial population are almost all incorrect—the winners of the two-armed bandit competitions would suggest (0 0 1 0 1), whereas the global optimum is (0 1 0 1 0). At the second stage, things are rather better, since the two-armed bandit would now suggest (0 0 0 1 0), which has only one bit incorrect. But how did the GA manage to achieve this on the basis of schema theory, when so many of the initial predictions were wrong? Secondly, the competitions are not independent: consider the initial population schema averages for $(*0***)$ and $(****1)$, which actually comprise *exactly the same* subset of $P(t)$.

Of course, this example is open to criticism as being based on a very simple problem, and using a very small population. However, as we shall see, it actually highlights quite neatly some of the problems of schema-processing arguments.

3.2. THE SCHEMA THEOREM

Table 3.1: Schema averages and fitness ratios for order-1 schemata in the populations of Tables 2.2 and 2.3

Schema	Population #1 average	Population #1 fitness ratio	Population #2 average	Population #2 fitness ratio
1****	1572	0.68	1551	0.56
0****	3455	1.49	3587	1.29
*1***	2100	0.90	2695	0.97
*0***	2475	1.06	2888	1.04
1	2903	1.25	2263	0.82
0	1458	0.63	3536	1.28
***1*	2200	0.95	4100	1.48
***0*	2513	1.08	2441	0.88
****1	2475	1.06	2320	0.84
****0	2100	0.90	3452	1.24

3.2.5 Does it work? Analysis

We can treat some cases in greater analytical depth. For example, consider a simple problem where strings have only two values:

$$f(x) = \begin{cases} \theta(>1) & \text{if } x \in S = (0****\cdots*) \\ 1 & \text{otherwise} \end{cases}$$

Suppose we apply a simple GA to this function—fitness-proportional selection, 1X and bit-wise mutation at a rate μ. For convenience, within this section we will denote the number of instances of S at generation t as $s(t)$. It is easily seen that crossover has no effect in this case, so that the number of instances of S in generation $t+1$ will be a binomial variable with parameters N (population size) and

$$p(t) = (1-\mu)\frac{\theta s(t)}{\theta s(t) + N - s(t)} + \mu \frac{N - s(t)}{\theta s(t) + N - s(t)}. \quad (3.5)$$

The mean number of instances of S at the next generation is thus

$$E[s(t+1)] = Np(t).$$

More generally, we can define a Markov chain on the states $\{0, 1, \ldots, N\}$, with binomial transition probabilities. Chapter 5 will look at Markov chains in some detail. For the moment, we merely note that the chain has the property that it is ergodic, since it is finite and all states communicate: the important thing about ergodicity for our purposes is that it means the process will have an equilibrium distribution. Without knowing a great deal about Markov chains, we can find this distribution simply from the intuitive idea that equilibrium will occur when the expected number of instances of S remains the same in successive generations. This implies that for t large enough,

$$\frac{s(t+1)}{N} = p(t)$$

and the fraction $p(t)$ will become a constant π. We thus obtain

$$\pi = (1-\mu)\frac{\theta\pi}{\theta\pi + (1-\pi)} + \mu\frac{1-\pi}{\theta\pi + (1-\pi)}.$$

Writing

$$\theta = 1 + \delta$$

reduces this to a quadratic

$$\delta\pi^2 + \pi[2\mu - \delta(1-\mu)] - \mu = 0.$$

Ignoring terms of $\mathcal{O}(\mu^2)$, this gives

$$\pi \approx 1 - \frac{\mu(1+\delta)}{\delta}.$$

This is clearly less than 1—i.e., the equilibrium state does not consist of strings all of which are members of S. For example, with $\theta = 1.1$ and $\mu = 0.02$, the equilibrium fraction of the population that belongs to S is only 78%. If we assume that similar results apply to all order-1 schemata, and further, that these schemata are independent, it is clear that even for moderate values of string length the chance of the population containing a string that represents *all* hyperplane competition winners is quite small. Thus the schema-processing theory tends to suggest arguments why a GA should *not* work efficiently!

3.3 Critiques of Schema-Based Theory

Our examples have suggested that all is not well with the 'classical' schema theory. Far from being special cases dreamt up to illustrate some unexpected but unimportant quirks of the theory, these examples are actually symptomatic of a wider critical literature.

3.3. CRITIQUES OF SCHEMA-BASED THEORY

Schema theory has been critiqued from several directions. Firstly, as Mühlenbein [173], has argued (echoing criticisms of the neo-Darwinian theory of evolution itself),

> ...the Schema Theorem is almost a tautology, only describing proportional selection...

Similarly, Holland records [127] that a criticism frequently levelled against it is that it is 'either a tautology or incorrect'. This seems a little unfair: if the theorem is incorrect, the mathematical errors should be pointed out—and nothing has appeared in the literature to challenge the mathematics. As for its being a tautology, there is of course a sense in which all mathematical theorems are tautologies—their conclusions are inherent in their premises, it is just that we did not originally discern them! Presumably, this criticism of the Schema Theorem supposes that the insights gained are so trivial as to be hardly worth the effort of pointing them out, but this is inevitably a subjective judgment.

However, Nick Radcliffe [198], Michael Vose [285] and others have raised a much more serious challenge, by pointing out that the Theorem extends to arbitrary subsets of the search space. In other words, there seems to be no special 'magic' about Holland's schemata. Vose [288] examined the function of mutation in a GA, and shown that a tiny change in the value of the mutation rate μ can cause a profound change in the GA's trajectory—a change which no schema analysis can possibly predict. Vose [294] contends further that the focus on the *number* of instances of a schema that survive is misplaced. What is far more important is *which* instances of a schema appear in the next population. The subset of strings represented by a schema usually has a fair amount of fitness variability (this is an aspect of epistasis), so the new set of strings in the next population is not necessarily as fit as the previous one. (On the other hand, if the epistasis is low, there are almost certainly simpler methods of finding the optimum anyway.)

All this is rather obscured by some statements of the Schema Theorem, including Holland's and Goldberg's, where the stochastic and dynamic nature of the quantities involved is brushed aside. In turn, this has led to the implications of the theorem being pushed further than can be sustained. One of the most glaring problems arises from the much quoted concept of 'exponentially increasing' trials of good schemata. It is worth rehearsing this argument for a moment, in order to point out where it goes wrong.

By ignoring the fact that the RHS of Equation (3.2) is an expectation, an inequality of the form

$$N(S, t+1) \geq (1+c)N(S,t)$$

can be set up for those schemata that are 'consistently above average', $c > 0$ expressing the 'excess fitness' required in accordance with Equation (3.4). This is then iterated to suggest that the number of instances of S will grow in accordance with a geometric progression (the discrete time analogy of exponential growth, which is more properly related to continuous time).

Why is this false? Firstly, we have the fact that the true relationship is stochastic, so we should not really confuse the actual number of instances of S in the equation by its expected value, unless we have an infinite population. But, more importantly, the 'growth factor' c will not remain at a constant level. Putting it another way, no schema *can* stay 'consistently above average'; rather, its fitness will shrink towards the population average. For if S *does* find itself with increased representation in the next population, the 'excess fitness' will almost certainly be smaller, since the intersection of S and $P(t+1)$ is larger. Thus there will be a tendency for the schema average to approach the population average, and at some point, because of the disruptive effects of crossover and mutation, the 'excess' will become a deficit. An example of this process can be seen in the previous section, where the schema average for $(0****)$ in Table 3.1 increased from 3455 to 3587, yet its fitness ratio has shrunk from 1.49 to 1.29. Finally, it should be unnecessary to point out that even if we had an exponential increase, it would have to end quite soon when we have a finite population. The rate of growth of a 'good' schema is much more likely to resemble a sigmoidal curve than an exponential. Nor is there any guarantee that the final state will consist of 100% representation of a given schema, as we saw in the example of Section 3.2.5 above.

In other words, Holland's Schema Theorem is a neat device for expressing what happens (on the average) to a schema S as we pass from one generation to the next. It is true that it makes a statement about S that does not rely on knowledge of any other schema. However, unless we know what happens to those other schemata as well, we cannot use it to extrapolate the trajectory of S beyond that next generation.

The idea of implicit parallelism is also of strictly limited application. It merely describes the number of schemata that are likely to be present given certain assumptions about string length, population size and (most importantly) the way in which the population has been generated—the last assumption being one that is unlikely to be true except at a very early stage of the search. Bertoni and Dorigo [19] have pointed out that Holland's $\mathcal{O}(N^3)$ lower bound relies upon a particular relationship between population size and m—the maximum schema order considered in the derivation, the assumption being that $N \sim 2^m$. By considering the case $N \sim 2^{\beta m}$ for a

3.3. CRITIQUES OF SCHEMA-BASED THEORY

range of values of β, they extended the argument to deal with an arbitrary population size, and show that the relationship can in fact be anything at all!

Moreover, the assumption that the population is independently and uniformly distributed is still required, so the lower bound may be of little practical relevance except in the case of the initial population. Finally, the idea that all these schemata can then be usefully *processed* in parallel is false for another reason, which ties in to problems with the two-armed bandit analogy.

A problem with an inappropriate application of the central limit theorem in Holland's original proof was highlighted in the 2nd edition of his book [126], although the essential core of his two-armed bandit argument was untouched. However, Rudolph [246] has recently shown that the two-armed bandit argument breaks down in a much more fundamental way: Holland's 'optimal' strategy for the bandit problem is comprehensively outperformed by an alternative approach that uses a sequential likelihood ratio test. Macready and Wolpert [155], as well as pointing out a flaw in Holland's mathematics, have independently argued, using a Bayesian framework, that there is no reason to believe that Holland's strategy is an optimal one. It follows that, even if we were to accept that the GA adopts a strategy of 'exponentially increasing trials', it is in any case *not* the best way of solving a competition between two competing schemata or hyperplanes.

Moreover, as mentioned in [218] (to be expanded upon later in Chapter 8), the assumption that hyperplane competitions can be isolated and 'solved' independently is false, unless the population has a very particular composition. That is, the GA is not really solving, concurrently and *independently*, $\mathcal{O}(N^3)$ two-armed (let alone multi-armed) bandit problems, since the arms cannot be 'moved' independently. Again, this has been illustrated in the example of Section 3.2.4.

In general, it may be difficult even to identify the arms properly; as shown later (Chapter 8), we may *think* we are contrasting $(1****)$ with $(0****)$, when the results we obtain are due to differences between (say) the schemata $(*00**), (*01**), (*10**)$ and $(*11**)$, an effect that statisticians are well acquainted with as 'aliasing' or 'confounding'.

Some of the other principles erected on the foundations of schema theory have also inevitably faced criticism. The PMA has been questioned: Antonisse [9] pointed out that what is meant by a 'schema' for higher-cardinality alphabets is not necessarily the same as it is in the binary case, where each * can only be replaced either by 0 or 1. With non-binary encodings, he argued, we should consider all subsets that could replace a *, counting each of them as a possible schema. From this perspective, there are many *more* schemata available to higher-cardinality alphabets. Of course the question is whether it matters—is the potential number of schemata important anyway, if they cannot be processed independently? On the other side of the argument, some practical support for the PMA does come from the fact that it may allow the use of smaller populations, as discussed in Chapter 2, and also from some experimental work by Rees and Koehler [212].

The BBH has also come in for some examination. Goldberg [97] introduced the idea of a 'deceptive' function—one which would lead the GA in the 'wrong' direction, since the 'building blocks' needed for a successful optimization were not present. As expected, GAs can find such problems very hard to solve [302], although the extent of the difficulty also depends on the amount of deception [107]. This is however mainly 'negative' evidence for the BBH; Mitchell *et al.* [169] described some experiences with the so-called 'Royal Road' function—a function that was meant to demonstrate the positive side of the BBH, in that it provided an environment for a GA to build an optimizing solution step by step, in the manner in which a GA was thought to behave.[2] It was somewhat disconcerting to find that the GA was comprehensively outperformed by a simple hill-climber—on a function that should have been ideal for the supposed *modus operandi* of a genetic algorithm.

However, that a GA really works like this is not likely anyway. Thornton [279] has pointed out that the requirements of the BBH are actually at variance with those of the Schema Theorem. The latter favours short, low-order schemata—as the building blocks become longer or more specific (as the BBH proposes) their representation in the next population is expected to diminish rather than increase. So there are many reasons why schema theory does not deliver the explanation for GA behaviour that it originally appeared to promise.

3.4 Surveying the Wreckage

In the late 1980s and early 1990s, schema theory and implicit parallelism seemed to provide a satisfying explanation for the utility of GAs, and even in some quarters for claims that they were in some sense the 'best' algorithm possible. 'Violations' of schema theory were thought to be significant: consider the statement

---

[2]Royal Road functions are described in Appendix A.

3.4. SURVEYING THE WRECKAGE

> *The most serious objection to ranking is that it violates the schema theorem...* [301]

where the objection was answered, not by a critical examination of the Theorem, but by re-interpreting the idea of fitness so that it could still be applied.

It should be appreciated that Holland's original account of the Schema Theorem and its ramifications were always rather more nuanced than that provided by some of the later enthusiasts for GAs. His discussion of 'exponential growth' in the second edition of his book [126], for example, clearly qualifies it as 'until its instances constitute a significant fraction of the total population' (although he leaves unstated what we should understand by 'significant'). Furthermore, Holland's emphasis was much less on finding optimal solutions, which assume a 'best-so-far' criterion in evaluating the effectiveness of a search, than it was on populations adapting to an environment, for which 'online average fitness' (the average of all individuals generated over the course of time) is arguably a more sensible measure of effectiveness. De Jong (who, ironically, did a great deal towards developing the optimizing aspects of GAs) has cautioned in a number of papers [56, 58] that we should not treat Holland's original vision of a GA (and its concomitant 'theory') merely as a function optimizer—preferring the acronym 'GAFO' for the application that has now become the most popular.

Nevertheless, the implications of these *caveats* have often been neglected by later researchers, so it may seem somewhat disconcerting to find that, over the course of the 1990s, Holland's GA theory has been so comprehensively eviscerated. Vose, for example, has said that

> *...schema theory tells us almost nothing about GA behaviour...* [293]

a judgment which clearly dismays John Holland [127]. It is not in any spirit of triumphalism that we have described the shortcomings of schema theory, nor would we wish to diminish the contribution of earlier researchers. This is the way science develops.

Neither is it to say that schemata must be completely consigned to the dustbin while we cast around for newer and better ways of describing the process of the genetic algorithm. The suggestion that schema theory tells us 'almost nothing' might be considered excessive, although the theory is clearly a lot less useful than we once thought. The *idea* of a schema is still a useful one, if rather more limited than it at first appeared. The notion that we should look for and try to exploit structure in a fitness function is surely sensible—although Holland's schemata will not necessarily accomplish that objective. Radcliffe [198, 199] has argued that in implementing a GA, we need to take account of the properties of the operators employed as well as the way in which a solution is represented. This is clarified by the work of Altenberg in putting the Schema Theorem into a wider context.

3.4.1 Price's theorem

Altenberg's work [5] related the Schema Theorem to earlier work in population genetics. Price [192] had shown how the expected value of some population trait could be related to the correlation of that trait with fitness. If we have some measurement function $F(x,t)$ for the trait (where x is a string in the current population), then Price's theorem can be expressed as follows:

Theorem 3.5 (Price) *For any pair of parents (y,z) at time t, define $\phi(y,z,t)$ to be the value of F averaged over all possible offspring of y and z. Then the average value of F in the next generation is*

$$E[F(x,t+1)] = \bar{\phi}(t) + Cov(\phi(t), r(y,t)r(z,t))$$

where $r(y,t)$ (resp. $r(z,t)$) is the fitness ratio of string y (resp. z), and $\bar{\phi}$ and $Cov(\cdot,\cdot)$ are respectively the mean of ϕ and its covariance with fitness, evaluated over all pairs (y,z) in each case. [These quantities are all parameterized by t to emphasize that they are calculated with respect to the current population.]

This looks fairly complicated, but as Altenberg shows, it is a more general formulation than the Schema Theorem. In particular, if the measurement function is given as

$$F_S(x,t) = \begin{cases} 1 & \text{if } x \in S \\ 0 & \text{if } x \notin S \end{cases},$$

so that the trait of interest is membership of schema S, the Schema Theorem emerges as a special case. With this definition of F, firstly we obtain

$$E[N(S,t+1)] = \bar{\phi}(S,t) + Cov(\phi(S,t), r(y,t)r(z,t))$$

where ϕ is written in terms of S to emphasize that it now represents those offspring of y and z which are members of schema S. Further, a lower bound on ϕ can easily be written for any S if we know the operator (e.g., 1X), and the schema theorem emerges in the form

$$E[N(S,t+1)] \geq \{1 - \eta(S,t)\}\{N(S,t) + NCov(F_S(y,t), r(y,t))\}.$$

It can thus be seen that for S to grow in frequency, there must be a positive correlation between the fitness of a string, and its membership of S. However, whether fitness and schema membership are correlated will be problem-dependent, and even when such a correlation exists it can only be measured with respect to the current population. Furthermore, while the Schema Theorem has very little to say about the operators—relegating them to transmission error status as in Equation (3.3)—Altenberg shows their relevance. For example, different recombination operators induce different fitness distributions, and 'progress' will only be made where the correlation is in the right direction.

Other work has also suggested the importance of correlations: Manderick et al. [157], for example, proposed that local parent-offspring correlations should be monitored in order to verify that the GA makes progress. While this idea is intuitively sensible, Altenberg produces two counter-examples to show that such measures are not sufficient to predict GA performance.

3.4.2 Formae and respect

Altenberg's is not the only work to point out the mistake of ignoring the constructive effect of the operators. Whereas a schema defines subsets of chromosomes that are similar in the specific sense that they have the same alleles at some stated loci, Radcliffe's concept [198, 199, 200, 202, 204] of a *forma* defines subsets of chromosomes that are similar in a more general way—perhaps in terms of the phenotype rather than in terms only of the genotype. For example, in the case of bin packing, the defining characteristic of a forma might be that a particular group of objects is packed together, while for the TSP, it might be that a chromosome possesses a particular edge. With Radcliffe's approach, it is possible to design operators *a priori* to have desirable properties in the context of the chosen means of representation. For example, he argues that it is often sensible to require operators to *respect* the formae of which the parents are instances. This means that where parents share a particular characteristic, their offspring should inherit this characteristic.

This aspect of *commonality* has been stressed several times; Reeves [214] pointed out the relevance of *which* string (out of all those in the common subset represented by the schema) is selected. Höhn and Reeves [121] suggested that the two-fold effect of crossover should be explicitly recognized—the first effect being the inheritance of common genetic material, the second being to explore the consequently reduced search space by means of an operator. Superior results for graph partitioning were obtained when this approach was used. This is also true for Falkenauer's 'grouping GA' [76], which uses the same idea, if not the same terminology. A 'commonality-based' Schema Theorem has been provided more recently by Chen and Smith [37].

3.4.3 Modifications

The acknowledged inadequacies of the original Schema Theorem have also stimulated new lines of enquiry. One is to take account of both 'gains' and 'losses' in the process of recombination, so that the Schema Theorem can be written as an equation, instead of a lower bound. The version in [37] is a recent attempt to deal with this question, but such ideas go back at least to Bridges and Goldberg [31].

Another approach has been taken by Stephens and Waelbroeck [272, 273] who construct an exact version of the Schema Theorem by recursively partitioning schemata into 'left' and 'right' pairs of lower order and shorter length than the original, until all schemata are of order 1. (The formulation is necessarily somewhat complicated, but it is closely related to that due to Altenberg [5], discussed above as Theorem 3.5.) Their analysis is able to demonstrate that the lower bound obtained from the traditional Schema Theorem can sometimes be far from sharp: schemata that (using the traditional approach) ought not to survive may in fact do so quite easily if the current population is able to create more instances than are lost by means of disruptive effects.

More recently Poli [188] extended this idea of recursively partitioning schemata in order to derive a probability distribution for the number of instances of a schema that survive. Poli points out that, given a *transmission probability* $\alpha(S,t)$ for schema S, the number of instances of S at $t+1$ is a Binomial variable with probability

$$\binom{N}{k}[\alpha(S,t)]^k[1-\alpha(S,t)]^{N-k}$$

and expectation
$$E[N(S,t+1)] = N\alpha(S,t).$$

The Stephens-Waelbroeck theorem can then be recast as a recursive formula for $\alpha(S,t)$. For example, if $\ell = 3$, and we ignore mutation, the transmission probability for $S = (*\,1\,1)$ is found by considering S itself and its 'components' $(*\,1\,*)$ and $(*\,*\,1)$:

$$\alpha(*\,1\,1,t) =$$

3.4. SURVEYING THE WRECKAGE

$$(1 - \frac{\chi}{2})\frac{N(*11,t)}{N}r(*11,t)$$
$$+ \frac{\chi}{2}\frac{N(*1*,t)}{N}r(*1*,t)\frac{N(**1,t)}{N}r(**1,t).$$

The difficulty in extending this is firstly the large number of recursive terms that are needed for high-order schemata, and secondly the complexities of calculating the required binomial probabilities. The latter problem Poli deals with by using well-known statistical inequalities to obtain stochastic lower bounds. However, despite using bounds that are fairly weak, it still involves a Herculean amount of labour in order to reach a statement of practical significance.

3.4.4 New test functions

Another viewpoint is that taken by John Holland himself [127]. He considers that the problem with the building block theory is the stochastic effects of a GA that are neglected in his (and others') use of the Schema Theorem. The original Royal Road (RR) functions [169] failed to demonstrate the utility of GAs for several reasons. As Watson et al. [298] have also pointed out, RR functions were too easy; this led in [298] to the development of 'hierarchical if-and-only-if' (HIFF) functions,[3] and in Holland's case [127] to 'hyperplane-defined functions' (HDFs), both of which are more challenging to search algorithms.

More fundamentally, RR functions highlighted a general problem in GAs—the phenomenon of 'hitchhiking'. This occurs where a particular low-order schema (perhaps just a single allele) has a significant influence on fitness. As selection and reproduction spread instances of this schema through the population, it is probable that alleles present in the first examples of the schema will hitch a ride, whether or not they are really important or not. For example, if $(1*1*\cdots*)$ has a high relative fitness, and its first occurrence is as $(101*\cdots*)$, there is a high probability that the population will soon consist largely of instances not only of $(1*1*\cdots*)$, but also of $(*0**\cdots*)$. This would not matter too much if $(1*1*\cdots*)$ really is important from a global perspective, but if (say) $(*11*\cdots*)$ is even better, the prevalence of $(*0**\cdots*)$ may prevent the GA from finding it. Of course, this is just another aspect of the problem of loss of diversity that we discussed in Chapter 2. Higher mutation rates or lower selection pressure might counteract it, but the consequently greater stochastic variability might prevent anything from becoming established at all.

---
[3]Further details of HIFF functions can be found in Appendix A.

This dilemma led Holland to propose 'cohort GAs' (cGAs) [127], where instead of the relative fitness of a string determining the *number* of offspring it can have, it determines the *frequency* with which the string is allowed to reproduce. This is accomplished by delaying reproduction by an amount that depends on fitness. However, although Holland motivates the cGA by schema-processing arguments, it is not clear that such arguments amount to a *sine qua non* for this approach.

3.5 Exact Models

Perhaps the most satisfying and elegant development stemming from dissatisfaction with schema-processing arguments is the formal executable model proposed by Vose [287] and Whitley [306], building on an earlier suggestion by Vose and Liepins [286]. This places the focus on the strings rather than on the schemata, although of course it is conceptually simple to move from one to the other. The complete dynamical system model of which this is the basis will be treated at length in Chapter 6, but it is appropriate to provide a simple introduction at this point.

A model that takes account of construction of strings (or schemata) as well as destruction needs to consider all possible ways of performing crossover and mutation. In the most general case—uniform crossover on complementary strings—*every* string in the Universe can potentially be constructed. This effect can be captured by a 3-dimensional array of probabilities, where the strings or vectors are indexed from 0 to $|\mathcal{A}|^\ell - 1$. The $(i, j, k)th$ entry of this array gives the probability that vectors i and j will recombine to produce vector k. The effect of mutation is similar to represent, but since only one string is involved, it is rather easier. Conceptually, this approach has proved to be rather successful in providing insights into GAs, as Chapter 6 will show.

However, for practical purposes, even a 10-bit chromosome would give rise to more than 10^9 entries, placing extreme demands on computational resources for applications of any reasonable size. Fortunately, as Vose and Liepins pointed out [286], a 2-dimensional matrix is sufficient to represent the effects of recombination. Assuming we have such a matrix for generating one arbitrary vector (suppose, for convenience, it is the all-zeroes vector, denoted simply by **0**), a simple permutation mapping suffices to determine the matrix for any other vector. Suppose we have the 'mixing matrix' with entries

$$M_{i,j}(0) = \mathbf{Pr}[\text{vectors } i \text{ and } j \text{ generate vector } \mathbf{0}],$$

3.5. EXACT MODELS

and we wish to generate the matrix for another vector k; this matrix is

$$M_{i,j}(k) = M_{i \oplus k, j \oplus k}(0)$$

where, as usual, $\oplus$ is 'exclusive OR', or addition modulo 2. (Note that we regard i, j, k as both indices and vectors for simplicity of notation.) As for the entries themselves, these are clearly operator-specific. For example, suppose we have UX with parameter p—the probability of selecting from vector i; the following is the mixing matrix for vectors of length 3, indexed in standard binary integer order.

$$M = \begin{bmatrix} 1 & p & p & p^2 & p & p^2 & p^2 & p^3 \\ q & 0 & pq & 0 & pq & 0 & p^2q & 0 \\ q & pq & 0 & 0 & pq & p^2q & 0 & 0 \\ q^2 & 0 & 0 & 0 & pq^2 & 0 & 0 & 0 \\ q & pq & pq & p^2q & 0 & 0 & 0 & 0 \\ q^2 & 0 & pq^2 & 0 & 0 & 0 & 0 & 0 \\ q^2 & pq^2 & 0 & 0 & 0 & 0 & 0 & 0 \\ q^3 & 0 & 0 & 0 & 0 & 0 & 0 & 0 \end{bmatrix}$$

To illustrate the permutation mapping, suppose we want the mixing probability for vectors $i = (1\,0\,0)$ and $j = (0\,1\,1)$ to produce vector $k = (1\,1\,0)$. This probability is clearly p^2q, since the second bit must come from j, and the others from i. In terms of the mapping function, we need

$$i \oplus k = 0\,1\,0 \quad \text{and} \quad j \oplus k = 1\,0\,1$$

and $M_{010,101}(0) = p^2q$ as expected. Whitley [306] gives an algorithm for generating the mixing matrix for some other operators in the case of binary vectors, and Whitley and Yoo [309] extend this to the case of permutation operators such as PMX. Mutation can also be handled in a similar fashion.

Having obtained a mixing matrix M, the model determines the evolution of the probability of vector 0 in the population as

$$\Pr[0, t+1] = s^T(t) M s(t) \tag{3.6}$$

where the vector $s(t)$ contains the probabilities of selection for all strings i, which implicitly includes the necessary fitness information. By use of the permutation mapping, Equation (3.6) can then be extended to all strings. Chapter 6 will show how these ideas can be developed in much greater detail.

Table 3.2: Goldberg's 3-bit deceptive function

String	Fitness
0 0 0	7
0 0 1	5
0 1 0	5
0 1 1	0
1 0 0	3
1 0 1	0
1 1 0	0
1 1 1	8

3.6 Walsh Transforms and deception

Another enduring legacy of the interest in schemata has been the application of Walsh transforms to binary GAs. Much of the analysis that we shall describe later in this book will require some understanding of Walsh analysis, and this seems an appropriate place to introduce the topic.

The idea of applying Walsh transforms in GAs was first developed by Bethke [20], but it is David Goldberg [98, 99] who deserves much of the credit for its wider recognition. The focus in these initial investigations was still on schemata: in particular, what sort of function it might be that could 'fool' a GA—in the sense that the schema averages would point the GA in the 'wrong' direction, away from the global optimum. Such a function is shown in Table 3.2. Note that definitions of this function in the literature may differ in unimportant details.

If we examine the schema averages here we find that while (1 1 1) is the optimal point, any schema that contains 1s is less fit than the corresponding schema that contains 0s: for example, $f(* * 1) < f(* * 0)$. Goldberg called this phenomenon *deception*. But how could such a function be constructed? This is where the Walsh transform is useful.

3.6.1 Function Decomposition

Suppose we have a function f defined for a variable x. It is natural to ask if this function can be decomposed into a superposition of simpler functions

$$f(x) = \sum_j w_j \zeta_j(x)$$

3.6. WALSH TRANSFORMS AND DECEPTION

where the ζ_j ideally should have some meaning that can be related to properties of the function f. In the case of GAs, we normally have x represented as a vector (often a binary string), which we shall denote by $\boldsymbol{x}$, and its components by x_i. Occasionally it will be convenient to abuse our notation and write x or $\boldsymbol{x}$ as seems most appropriate. The most well-known set of functions for binary strings is the set of *Walsh* functions, which may be defined as follows:

$$\psi_j(\boldsymbol{x}) = \prod_{i=1}^{\ell}(1-2x_i)^{j_i}$$

where $\boldsymbol{j}$ is the binary vector representing the integer j (and j_i represents its components). There are several equivalent definitions: one such is

$$\psi_j(\boldsymbol{x}) = \varpi(\boldsymbol{x} \wedge \boldsymbol{j})$$

where $\wedge$ is the bitwise AND operator, and ϖ is the parity function

$$\varpi(\boldsymbol{y}) = \begin{cases} +1 & \text{if } \sum_i y_i \text{ is even} \\ -1 & \text{if } \sum_i y_i \text{ is odd} \end{cases}.$$

For example, if $\boldsymbol{x} = (1\,1\,0)$ and $j = 5$, which means that $\boldsymbol{j} = (1\,0\,1)$, we find that $\boldsymbol{x} \wedge \boldsymbol{j} = (1\,0\,0)$, with odd parity, so that

$$\psi_5(1\,1\,0) = -1.$$

For strings of length ℓ, there are 2^ℓ Walsh functions, so that the full decomposition is

$$f(\boldsymbol{x}) = \sum_{j=0}^{2^\ell-1} w_j \psi_j(\boldsymbol{x}).$$

Such a decomposition is of course a close analogy of what is done for functions of real numbers. There it is known as a Fourier decomposition, and entails the superposition of trigonometric functions—of which Walsh functions are a discrete analogue, forming a set of rectangular waveforms. For example, the function $\psi_1(x)$ is as shown in Figure 3.2.

The Walsh coefficients $\{w_j\}$ can be obtained from the values of $f(x)$ by means of the Walsh transform

$$w_j = \frac{1}{2^\ell} \sum_{x=0}^{2^\ell-1} f(x)\psi_j(x) \tag{3.7}$$

Note that this can also be expressed in a matrix form as

$$\boldsymbol{w} = \frac{1}{2^\ell}\widetilde{\boldsymbol{W}}\boldsymbol{f}$$

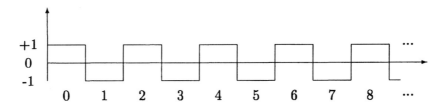

Figure 3.2: The Walsh function $\psi_1(x)$

where the matrix $\widetilde{W}$ has elements

$$\widetilde{W}_{i,j} = (-1)^{i^T j} \quad \text{for } i, j \in \{0, 1\}^\ell$$

and $i^T j = \sum_{k=0}^{\ell-1} i_k j_k$ is an *inner product*. As in the case of Fourier transforms, there are computational algorithms that can speed up the process. One such is Yates' algorithm [319] (see Exercise 6 at the end of this chapter), but a wealth of detail on such matters can be found in [17].[4]

All this is interesting, but what does it have to do with schemata? Readers may observe a strong similarity between Figures 3.2 and 3.1. This is no coincidence. With the use of some simple algebra and arithmetic it can be shown that schema averages are easily computed from appropriate subsets of the Walsh coefficients, and *vice-versa*. It can easily be seen that, for example,

$$w_0 = \bar{f}$$

the mean of all fitnesses, while

$$f(**0) = w_0 + w_1 \quad \text{and} \quad f(**1) = w_0 - w_1,$$

with similar results for more specific schemata. Table 3.3 provides the coefficients that result from such calculations for the deceptive function of Table 3.2.

It is thus possible to work backwards from the deception conditions $f(**1) < f(**0)$ etc., in order to obtain conditions on the Walsh coefficients that will deliver us a deceptive function. This is in fact how Goldberg's function was originally determined. (Two of the constraints derived in [98] are, however, incorrect; more will be said on this point in Chapter 8).

---

[4] It should be noted that the Walsh literature has different notions of numbering the Walsh functions. That used in this book follows Goldberg [98, 99] rather than the standard order used in [17].

3.6. WALSH TRANSFORMS AND DECEPTION

Table 3.3: Goldberg's 3-bit deceptive function: Walsh coefficients

Coeff	w_0	w_1	w_2	w_3	w_4	w_5	w_6	w_7
Value	3.50	0.25	0.25	1.00	0.75	1.50	1.50	-1.75

3.6.2 More on deception

Early theoretical work was still heavily influenced by schema theory, and tended to concentrate quite markedly on deception and its analysis. Whitley was able to characterize the degrees of deception in terms of different orders of schema competition [302, 305], and he also defined what is meant by the concepts 'fully' and 'consistently' deceptive. For binary strings, he also showed that the deceptive attractor is the complement of the global optimum, and although it is not necessarily a Hamming space local optimum, it must at least be a Hamming neighbour of such a local optimum. This is a convenient point to remark that discussion of what is a hard or easy problem for a GA is often bedevilled by a lack of caution in transferring between those ideas of relevance specifically to neighbourhood search and those relating to GAs. It cannot be assumed that what is a local optimum for one representation or algorithm is necessarily so for another. We shall return to this question in Chapters 8 and 9, but this will serve as an early warning that such analyses may need to be fairly sophisticated.

Whitley also pointed out that in order to be deceptive, it was sufficient that (assuming, without loss of generality, the global optimum to be the all-1s string) for $k = 1, \ldots, \ell - 2$, any string with k 1s is fitter than any string with $(k+1)$ 1s—a rather simpler means of construction than the Walsh-based approach. He also argued strongly that the only challenging problems for a GA are deceptive. However, Grefenstette [107] later produced an example to show that deception does not necessarily make life hard for a GA:

$$\begin{aligned} f(x) &= 2 & \text{if } x \in (1\,1\,1 * \cdots *) \\ &= 1 & \text{if } x \in (0 * * * \cdots *) \\ &= 0 & \text{otherwise} \end{aligned}$$

Since the schema $(1 * * \cdots *)$ has a mean fitness only half that of $(0 * * \cdots *)$, this function certainly qualifies as deceptive. Nevertheless, as Grefenstette discovered, most GAs will 'solve' this problem easily. The reason is that once an instance of $(1\,1\,1 * \cdots *)$ is found (which is not unlikely, given a reasonable population size), this particular subset of $(1 * * \cdots *)$ will quickly spread and other instances of the lower-order schema will disappear. The problem is deceptive from a global perspective, but in practice the GA will only 'see' the local situation. Once again, Vose's earlier point is relevant—it is the strings themselves that matter, not just the schemata. The strings in the population at a given time may have schema averages that are quite different from those that would be obtained from the Universe of all strings. However, while deception does not necessarily mean practical difficulty for a GA, it should not be assumed that its absence make life easy. The fact is that *any* estimates of (true) average schema fitness could be wildly inaccurate—particularly for low order schemata—unless populations are very large; even with large populations, the pressures of selection will fairly soon restrict the visible instances of a schema to a very small fraction of the subset as a whole. (Consider a first-order schema for binary strings of length $\ell = 50$, say. This schema comprises $2^{49} \sim 10^{14}$ strings. Very few GAs will see even 1% of these strings during a search!)

Finally, there has to be some doubt as to the relevance of schema fitness *averages* in any case, if we are interested in optimization. It is much more likely that each schema's *extrema* will be relevant, and the Schema Theorem says nothing about them. Later, in Chapter 8, we shall discuss some experimental evidence that extrema may be more important than averages.

3.7 Conclusion

In this chapter we have described and critiqued the traditional schema processing arguments for GAs, as developed initially by Holland. We have also considered some of the modifications and extensions that have been derived from this basic framework.

While the idea of a schema has some value in explaining GA performance, we are far from convinced that it holds the key to a complete understanding. There are many other aspects of genetic algorithms that have been investigated from a theoretical standpoint, and the schema concept is not always (or even often) relevant to some of these approaches. We shall discuss some of these ideas in subsequent chapters, but before we do so, there is an important matter to which we should pay some attention.

Some early accounts of GAs tended to stress their applicability to optimization in such a way as to imply that they were unequivocally the best approach. Partly this is done by somewhat nebulous appeals to 'nature' or 'evolution', as discussed in Chapter 1. However, some of the arguments have been based on the theory associated with Holland's schemata. The notion of implicit parallelism, for example, and the principle of minimal alphabets, have both been used to justify GAs as a way of maximising our chances of finding high-quality solutions. As observed in Chapter 1, John Holland himself has made such claims [125], and David Goldberg's book [96] does not avoid giving that impression. More than 10 years on from these works, it is still not uncommon to find claims that GAs are 'global optimizers' with the implication that they will find the global optimum for any problem. In some sense, this is true, as we shall discuss in Chapter 5, but this ignores practical questions of speed of convergence etc. What we certainly *cannot* say is that GAs are uniformly the best technique to apply for any problem. Probably nobody would really believe such a claim, but the publication of the so-called 'No Free Lunch' theorem [314] has demonstrated conclusively that such a claim is impossible. The issues raised by this theorem and its implications for optimization algorithms in general, and GAs in particular, will be outlined in the following chapter.

3.8 Bibliographic Notes

The basic ideas of schemata can be explored in greater depth by reading the books by David Goldberg [96] and John Holland [126]. Their ideas are reproduced more or less uncritically by books and tutorial papers of the early 1990s, although Melanie Mitchell's book [170] is partly an exception. However, readers who consult leading edge research papers of this era (such as those in the *Foundations of Genetic Algorithms* series [345]—[348]) will find an increasingly powerful critique of the building block hypothesis, implicit parallelism and so on.

In any consideration of the importance (or otherwise) of schemata and the historical development of schema-processing concepts, the papers by Lee Altenberg [5], Nick Radcliffe [200, 202, 205], Michael Vose [285, 288, 294] and Darrell Whitley [302, 305] are essential reading. More recent work has suggested a new synthesis where schemata find a place within a more comprehensive theory. The papers of Chris Stephens [272, 273] and Riccardo Poli [187, 188] are of considerable relevance in this re-appraisal.

Exercises

1. Consider the following population of 4 chromosomes

   C_1 11101111
   C_2 10101011
   C_3 00010100
   C_4 01000011

   and six schemata defined as

   S_1 1*******
   S_2 0*******
   S_3 ******11
   S_4 ***0*01*
   S_5 1*****1*
   S_6 1110****

   (a) State the order $k(S)$ and length $l(S)$ of each of these schemata.

   (b) How many instances of each of these schemata exist in the population?

   (c) Calculate the probability of disruption of each schema under bit-wise mutation with $\mu = 0.01$.

   (d) Calculate the probability of survival of each schema under one-point crossover.

2. A schema is given by $S = (1***1***)$. Calculate the probabilities that the offspring will still be a member of S for 1X and 2X assuming the other parent is not.

3. Repeat the above calculation for UX with $p = 0.1, 0.2, \ldots, 0.5$.

4. The 2-point crossover operator for a string of length ℓ is usually implemented by choosing (using a uniform probability distribution) 2 distinct crossover points x, y such that $1 \le x < y \le \ell - 1$. Consider a schema S of length s and order 2.

   (a) Find the probability of choosing crossover points (x, y) that do *not* disrupt S

   (b) Hence compare the disruptive potential of 2-point crossover with that of 1-point crossover.

5. Construct the mixing matrices for generating the vector **0** using 1X and 2X in the case of strings of length 3.

3.8. BIBLIOGRAPHIC NOTES

6. Yates' algorithm [319] is a simple procedure for finding Walsh coefficients. The function values are listed in a column in standard order. Then a second column is obtained by first adding successive pairs of values in the first column, then by subtracting the second member of each pair from the first. Thus, in the case $\ell = 3$, we obtain

f_0	$f_0 + f_1$
f_1	$f_2 + f_3$
f_2	$f_4 + f_5$
f_3	$f_6 + f_7$
f_4	$f_0 - f_1$
f_5	$f_2 - f_3$
f_6	$f_4 - f_5$
f_7	$f_6 - f_7$

   The same pattern is repeated to produce column $i+1$ from column i for $i = 3, \ldots, \ell + 1$. Dividing the last column by 2^ℓ gives the Walsh coefficients.

   (a) Verify that the coefficients in Table 3.3 are correct using this algorithm.

   (b) For $\ell = 3$, show that the pattern of signs in the last column is identical to that given by $\widetilde{\boldsymbol{W}}$.

Chapter 4

No Free Lunch for GAs

This chapter is something of a digression, and could be omitted by readers who are interested solely in the variety of theoretical investigations that have taken place in the development of genetic algorithms. However, that would be a pity, as the ideas associated with 'No Free Lunch' are helpful in clarifying many issues arising from the use of heuristic search methods in general, as well as having a particular bearing on GAs.

4.1 Introduction

Machine learning is an area of research that has overlapped with GAs in several ways, and it was from researchers in the machine learning community that the idea of 'No Free Lunch' originated. However, optimization is not directly under investigation in machine learning—rather it is the performance of classifiers that learn to predict output values from input data. The generally accepted measure of performance is the classifier's capability for *generalization*: the ability to predict outputs using inputs that were not part of the data on which the classifier learned. Schaffer [253] put forward a 'conservation law' that showed that in principle, the generalization capability of all classifiers is the same.

This immediately stimulated enquiry as to whether similar properties held for optimization, and the No Free Lunch (NFL) Theorem for optimization was proposed by Wolpert and Macready in 1995, and after stirring up a heated debate,[1] the paper was published in 1997 [314]. NFL is concerned

---

[1]Some of the points raised and answered in this debate can be found in the archives of the GA-List (especially volume 9) at http://www.aic.nrl.navy.mil/galist/

with the performance of algorithms that search for the optimum of a cost function.

Summarizing briefly, what Wolpert and Macready show is that, averaged over all possible functions, the performance of all search algorithms that satisfy certain conditions is the same. For example, if these conditions hold, random search is on average as good as any more sophisticated algorithm such as a GA. However, the idealized search algorithm they describe is different in some important respects from 'real' search algorithms such as simulated annealing, tabu search, or genetic algorithms. Nonetheless, it is frequently claimed that these results are relevant to such algorithms—as for example,

> ..., if simulated annealing outperforms genetic algorithms on some [sub]set ϕ [of the set of all functions $\mathcal{F}$], genetic algorithms must outperform simulated annealing on $\mathcal{F} \setminus \phi$. [314]

This is true in terms of what Wolpert and Macready mean by 'outperform', but as we shall argue below, their interpretation is somewhat idiosyncratic, and makes certain assumptions that may not be easily verified in practice. Firstly, we present the theorem, largely in the notation of [314], which is slightly different from that used in the rest of this book.

4.2 The Theorem

Assume we have a discrete search space $\mathcal{X}$, and a function

$$f : \mathcal{X} \mapsto \mathcal{Y} \subset \mathbb{R}.$$

For convenience, the value of $f(\mathbf{x})$ is called the cost of $\mathbf{x}$. The general problem is to find an optimal solution, i.e., a point $\mathbf{x}^* \in \mathcal{X}$ that minimizes or maximizes f.

Wolpert and Macready assume a search algorithm A that has generated a set of m *distinct* points $\{d_m^\mathbf{x}(i)\}$, and associated cost values $\{d_m^y(i)\}$ where $y = f(\mathbf{x})$ and the index $i = 1, \ldots, m$ implies some ordering (the most obvious one being with respect to the search chronology). For convenience, the whole ensemble of points and cost values can be denoted by d_m.

Thus initially, starting from a point $\mathbf{x}_1$ with cost y_1, we have $d_1^\mathbf{x}(1) = \mathbf{x}_1$ and $d_1^y(1) = y_1$. Subsequent points are generated by A based on d_m; that is, A is a function

$$A : d \mapsto \mathbf{x}, \text{ where } \mathbf{x} \in \mathcal{X} \setminus \{d_m^\mathbf{x}(i)\}.$$

4.2. THE THEOREM

The information generated by this sequence of points can be encapsulated in a frequency table or 'histogram', $\vec{c}$, of the cost values $\{d_m^y(i)\}$; the quality of the algorithm's performance can be measured in terms of some characteristic of $\vec{c}$ such as its minimum, mean, or median. For a given f, the quantity of interest is thus the conditional probability $P[\vec{c}|f, m, A]$. For this histogram, Wolpert and Macready prove the following:

Theorem 4.1 (No Free Lunch) *For any pair of algorithms A_1 and A_2,*

$$\sum_f P[\vec{c}|f, m, A_1] = \sum_f P[\vec{c}|f, m, A_2].$$

where the sum is over all possible functions f.

As originally stated, the theorem also assumes that A is deterministic—which search methods such as GAs and simulated annealing are definitely not. However, the definition of A can be extended so that the NFL theorem also encompasses stochastic algorithms.

More generally, attention can be shifted to some *performance measure* $\Phi(\vec{c})$. As a corollary, it is clear that the average of $P[\Phi(\vec{c})|f, m, A]$ over all functions f is independent of A, which provides the justification for such interpretations of the theorem as those quoted above.

4.2.1 Representations

Although [314] was the first rigorous statement of the 'No Free Lunch' concept, it is not the only one, and its proof is certainly not the easiest to follow. Radcliffe and Surry [206] approached this topic by considering the related issue of the choice of representation. Earlier (Chapter 2) we described the way in which an encoding function c is often used to represent the search space $\mathcal{V}$ indirectly in a representation space $\mathcal{X}$. It is intuitively obvious that some representations will work better than others in any particular case—for example, consider the experimental evidence regarding binary or Gray coding.

While early work in GAs tended to assume by default that the conventional integer-to-binary code was always appropriate, it does have problems. For example, suppose that the optimum of a function occurs at a point whose value is the integer 32, and that we are using a 6-bit chromosome, which means the binary mapped value is (1 0 0 0 0 0). If the function is reasonably smooth, a good practical approximation to this value is 31, which maps to the maximally different string (0 1 1 1 1 1). Conversely, the (very close) binary value (0 0 0 0 0 0) represents the integer 0, some distance away from the optimum! This difficulty led Caruana and Schaffer [33] to advocate the use of a Gray code instead. Technically, a Gray code defined on binary strings of length ℓ is a Hamiltonian path in the binary hypercube $\{0,1\}^\ell$. There are many such paths, and therefore many Gray codes (at least $2^\ell \ell!$), but the one that is commonly meant is the binary reflective Gray (BRG) code, which is easily obtained from the standard binary code as follows:

$$GB: \{0,1\}^\ell \to \{0,1\}^\ell \quad \begin{cases} x'_1 = x_1 \\ x'_i = x_{i-i} \oplus x_i, & i > 1 \end{cases}$$

The inverse mapping is given by

$$BG: \{0,1\}^\ell \to \{0,1\}^\ell \quad \begin{cases} x'_1 = x_1 \\ x'_i = x'_{i-i} \oplus x_i, & i > 1 \end{cases}$$

where, as usual, $\oplus$ is addition modulo 2. Table 4.1 shows what this looks like for the case $\ell = 3$. These mappings can also be stated in the form of matrices [310]. The GB matrix has 1s along the diagonal and upper minor diagonal, while the BG matrix has 1s along the diagonal and in the whole of the upper triangle.

Table 4.1: The binary reflected Gray code for $\ell = 3$. The 'reflected' character of the coding can be seen by inspecting the axis of symmetry inserted after the 4th line in the table. Disregarding the first bit, there is a clear reflection in this axis.

Binary	Integer	Gray (BRG)
0 0 0	0	0 0 0
0 0 1	1	0 0 1
0 1 0	2	0 1 1
0 1 1	3	0 1 0
1 0 0	4	1 1 0
1 0 1	5	1 1 1
1 1 0	6	1 0 1
1 1 1	7	1 0 0

Of course, this is not the only coding that could be chosen. A mapping can be permuted in many ways, producing a simple re-labelling of the elements. Table 4.2 shows a simple example.

Radcliffe and Surry used this fact in introducing the concept of *isomorphic* algorithms. Consider two search algorithms A and B, each of which

4.2. THE THEOREM

Table 4.2: The 24 (= 4!) encoding functions that could be used to map binary strings of length 2 to the integers $\{0, 1, 2, 3\}$. The 'standard' binary encoding function is the one in the top left corner. Immediately beneath it is the binary reflected Gray code.

s	x	s	x	s	x	s	x
0 0	0	0 0	1	0 0	2	0 0	3
0 1	1	0 1	0	0 1	0	0 0	0
1 0	2	1 0	2	1 0	1	1 0	1
1 1	3	1 1	3	1 1	3	1 1	2
0 0	0	0 0	1	0 0	2	0 0	3
0 1	1	0 1	0	0 1	0	0 0	0
1 0	3	1 0	3	1 0	3	1 0	2
1 1	2	1 1	2	1 1	1	1 1	1
0 0	0	0 0	1	0 0	2	0 0	3
0 1	2	0 1	2	0 1	1	0 0	1
1 0	1	1 0	0	1 0	0	1 0	0
1 1	3	1 1	3	1 1	3	1 1	2
0 0	0	0 0	1	0 0	2	0 0	3
0 1	2	0 1	2	0 1	1	0 0	1
1 0	3	1 0	3	1 0	3	1 0	2
1 1	1	1 1	0	1 1	0	1 1	0
0 0	0	0 0	1	0 0	2	0 0	3
0 1	3	0 1	3	0 1	3	0 0	2
1 0	1	1 0	0	1 0	0	1 0	0
1 1	2	1 1	2	1 1	1	1 1	1
0 0	0	0 0	1	0 0	2	0 0	3
0 1	3	0 1	3	0 1	3	0 0	2
1 0	2	1 0	2	1 0	1	1 0	1
1 1	1	1 1	0	1 1	0	1 1	0

produces a sequence of distinct points in the search space $\mathcal{V}$. For A and B to be isomorphic, we need to have a permutation of the representation space $\mathcal{X}$ that maps one sequence to the other. Once again, there is the requirement for the points visited to be distinct, i.e., there is no re-visiting of previous points in the search space. Assuming this to be the case, it is then fairly easy to show that

1. If A and B relate to the same objective function and the same representation space $\mathcal{X}$, they are isomorphic.

2. Over all encoding functions represented by a permutation of $\mathcal{X}$, average performance is the same.

Whereas Wolpert and Macready's NFL averages across all functions f, Radcliffe and Surry show that NFL also applies to encodings of individual functions, and 'the' No Free Lunch Theorem emerges as a consequence.

4.2.2 Gray versus binary

One special case that has been investigated at some length is the comparison of Gray and binary coding. Table 4.3 illustrates what can happen. Here we have a simple function whose domain is the integers from 0 to 7. The range of the function is also the integers 0 to 7. (Limiting the domain in this way may seem excessively restrictive, but in cases where the domain is discretized for the purpose of binary coding, this *is* effectively the domain of the function. The assumption that the range consists of distinct well-ordered values is perhaps more of a problem, but this is only an illustration.)

The function in the table is just one of $(2^3)! = 40320$ possible functions that could be constructed, but it has an interesting property: if we define a natural neighbourhood over the integers—i.e., $(x-1)$ and $(x+1)$ are the natural neighbours of x—there are 4 maxima, occurring at the odd values of x. (We can consider the neighbourhood to 'wrap around' the endpoints of the domain if we wish. In this case it does not matter whether it does or not.) This is clearly the greatest number of maxima possible for a function defined over just 8 points. Now consider what happens with a bit-flip neighbourhood for the corresponding binary and Gray codings. In the binary case, every point has a better neighbour except (1 1 1), so there is only one maximum. However, for the Gray case, the points (0 1 0), (1 1 1) and (1 0 0) are all maxima with respect to their neighbours.

What of the other 40319 functions? Rana and Whitley [207] have enumerated them and counted the number of optima under each form of representation. The results were interesting, and are summarised in Table 4.4.

4.2. THE THEOREM

Table 4.3: An example of the effect of different coding mechanisms. The function f is one of the $(2^3)!$ possible mappings of the integers x to fitness values. However, with a standard neighbourhood operator, there are different numbers of optima for the 3 ways of representing x.

x	f	b	g
0	0	0 0 0	0 0 0
1	2	0 0 1	0 0 1
2	1	0 1 0	0 1 1
3	4	0 1 1	0 1 0
4	3	1 0 0	1 1 0
5	6	1 0 1	1 1 1
6	5	1 1 0	1 0 1
7	7	1 1 1	1 0 0

Table 4.4: A head-to-head comparison of Gray code and standard binary code for each case of the 40320 functions as defined in the text. k is the number of optima given by the natural integer neighbourhood. These statistics were originally reported in [311].

k	frequency	wins	loses	ties
1	512	448	0	64
2	14592	6384	2176	6032
3	23040	7088	6704	9248
4	2176	0	2160	16
Totals	40320	13920	11640	15360

In this table, the number of cases where Gray code 'wins', 'loses' or 'ties' (i.e., has fewer than, more than or the same number of optima as the standard binary code for the same function) is recorded against the actual number of optima generated by the natural integer neighbourhood. While the total number of optima was the same in each case (in accordance with NFL), Gray code beats binary on a head-to-head comparison: i.e., it is better more often, even if not by much. The flip side of this is that when it is worse than binary, it is occasionally much worse. Of particular importance is the fourth line of the table, which shows that it is precisely on the 'hardest'[2] functions where Gray loses its advantage. The example in Table 4.3 is one case of this phenomenon.

Subsequently Whitley [311] was able to show that this type of behaviour is not limited to $\ell = 3$. By means of an inductive argument, he showed (subject to a slight clarification discussed in [312]) that if we limit the class of functions of the type explored above to those with the *maximum possible* number of optima for the natural integer neighbourhood, the Gray code always induces *more* optima than the corresponding binary code. As a corollary of this, it is clear from NFL that on average Gray coding must 'win' over all other functions. This is not only an impressive piece of theory, but it also bolsters the argument in practice for using Gray codes as suggested by [33], at least if we believe (as seems reasonable) that the cases where Gray coding loses are unlikely to be representative of real-life problems.

4.2.3 Adversary Arguments

Culberson [42] has an interesting viewpoint on NFL, and his explanation is perhaps the most straightforward way of thinking about these issue. He imagines an 'adversary' sitting in a box, receiving strings from the algorithm for evaluation, and reporting back a fitness value. In the general case, the adversary is completely unrestricted and passes back any valid value. It then follows that the NFL table $\vec{c}$ of cost values generated after a given number of requests is simply a random selection of valid values and for any performance measure based on $\vec{c}$, no algorithm can do any better than random search. Putting it another way, whatever criterion the algorithm chooses to construct its 'next' string, the adversary remains indifferent. It should be noted that indifference is the relevant characteristic of the adversary—it

---

[2] Deciding what is hard or easy is difficult, as we shall see in Chapters 8 and 9. Here we are making the assumption that the more optima a function has, the harder it is. While this is intuitive, we should realize that the *number* of optima is not the only important factor.

is not necessary to attribute any malevolence to the adversary, despite the ordinary connotations of the word. English [69] has criticised Culberson's argument on the grounds that such malevolence is implied, but it is by no means clear that this is really inherent in the argument.

4.2.4 Information-theoretic arguments

More recently English [68, 69, 70] has applied information-theoretic ideas to argue that in general information is conserved. Of course, English is using the term, not in the everyday sense, but in the technical sense associated with Shannon, in the case of statistical information, or Chaitin, for algorithmic information. By conservation he means that an algorithm's good performance on some problem instances is paid for by poor performance on others. More will be said on this aspect in the next section.

4.3 Developments

Several points have been raised in response to the NFL theorem. The concept of 'all possible functions' is necessary for the original NFL arguments, but what use is this as a practical guide? English [68] extends NFL so that it does not need to rely on a notion of how functions are (statistically) distributed. Almost all *individual* functions are highly random ones [68], in the sense that there is no compressed description of the function that is much shorter than a list of the function's permissible arguments and their associated values. Different optimizers simply sample these values in different sequences. The simplest approach would be to have a random sequence, which does in fact guarantee to find a solution within the best α of all solutions (as a fraction of the Universe) with a probability that depends only on α and the number of samples taken.

This is not hard to see: suppose we rank the Universe of solutions in ascending order of fitness, denoting the list by $x_{[1]}, x_{[2]}, \ldots, x_{[K]}$ where $K = |\mathcal{X}|$ is the cardinality of the search space. We define the set of élite solutions $\mathcal{B}$ as

$$\mathcal{B} = \bigcup_{i=1}^{B} x_{[i]}$$

where $B = \lceil \alpha K \rceil$ for some fraction α. Now if we sample randomly (without replacement) n times, it is clear that the probability that at least one of the points sampled is in this élite set is

$$P = 1 - (1 - \alpha)^n$$

so that in order to achieve a particular value of P, we need

$$n = \lceil \log(1-P)/\log(1-\alpha) \rceil.$$

For example, with $\alpha = 10^{-5}$ and $P = 99.99\%$, we need fewer than 1 million samples (actually $n = 921029$). In practice, sampling *with* replacement is easier to implement, so that a specific n does not generate n unique points. However, given the size of the search spaces usually encountered in COPs, the difference is not likely to be great.

English [68] uses his information-theoretic argument to show that random search is not anything special: almost all optimizers will find such 'good' solutions with few samples (relative to the size of the search space), and in this sense, optimization is not a hard problem. (It is interesting, although not directly relevant to this discussion, that English's formulation allows a unified treatment of optimization and learning as different aspects of function *exploration*, and that a corresponding analysis of learning shows that it is a hard problem [69]. Intuitively, we can argue that for optimization, if our algorithm doesn't work, we can always retreat to the position of using random search. However, in the case of learning we have no such fall-back position.)

But this raises the question why we use GAs (or any other metaheuristic) at all, instead of just using random search? Firstly, the fitness distributions of COPs may have a very long tail, and although we may find a point in the élite *group* relatively easily, this is no guarantee that its objective function *value* is within a useful fraction of that of the optimum. Secondly, in choosing to use GAs (say), the implied assumption must be that there is something 'structural' in the problems whose solutions are sought by engineers, operations researchers, statisticians etc. This implicit reasoning must also be a factor in choosing Gray codes over binary, as noted in Section 4.2.2. Practitioners generally seem to believe that there is something about 'their' problems that make them amenable to solution by a particular method (such as a GA) more rapidly and efficiently than by random search. What this 'something' is, is hard to define rigorously, but in [70] English provides some clues when he shows that low-order polynomials do not satisfy the conditions for conservation of performance.

4.3.1 Free appetizers?

Droste *et al.*[63] have also considered the possibility of restricting the complexity of functions in some way. Exactly how we recognize a restriction is an interesting question that we shall explore further in Chapter 8; in [63]

4.3. DEVELOPMENTS

several scenarios are presented, which it is argued are more representative of real-world optimization problems. By enumerating a toy problem, it can be demonstrated that different evolutionary algorithms do indeed have different performance, although in the case considered in [63] the differences were small. Hoping for a free lunch may be a little optimistic, but some algorithms may at least gain a free appetizer!

In a further paper [64], the same authors present an 'Almost No Free Lunch' (ANFL) theorem. While there may be a free appetizer for an algorithm, in general the gains are small even in the restricted case: they show that for every (restricted) scenario where a given algorithm has an advantage, it is possible to construct many related functions where it does not.

4.3.2 Permutations

However, Igel and Toussaint [132] have recently provided a more hopeful interpretation of NFL. They build on work by Schumacher *et al.*[259], who showed that we can obtain a NFL result for *any subset* F of the set of all possible functions, if and only if F is closed under permutation (*c.u.p.*). Several of the NFL results have focused on the transformation of one function into another one by means of a permutation, i.e., given a permutation $\pi : \mathcal{X} \to \mathcal{X}$, we can produce a function $f' = f \circ \pi$ from any $f \in F$. Requiring the set F to be *c.u.p.* implies that every such f' must also be in F.

But what sort of classes of functions are *c.u.p.*? Igel and Toussaint show firstly that the absolute number of such classes of functions grows extremely fast—the number of (non-empty) subsets of $\mathcal{F}$ is $2^{\mathcal{N}} - 1$ where

$$\mathcal{N} = \binom{|\mathcal{X}| + |\mathcal{Y}| - 1}{|\mathcal{X}|}.$$

However, this is still a tiny (and decreasing) fraction of $\mathcal{F}$; might it be that the 'interesting' functions are predominantly in the rest of the space? It would certainly be very convenient if this were so! Happily, it appears this may indeed be the case: Igel and Toussaint show that by imposing a neighbourhood structure on the search space and function—creating what we shall later call (Chapter 9) a *landscape*—the *c.u.p.* property is likely to be broken, unless some rather severe constraints also hold. For example, if the number of optima induced on this landscape is less than maximal for all functions in a subset F, then F cannot be *c.u.p.*, and therefore a NFL result cannot hold. (Note that this is in agreement with Whitley's analysis of the relative merits of Gray and binary codes.) If we believe—as seems reasonable, on the empirical evidence—that 'real' functions are not as difficult as they could possibly be (e.g., by having the maximal number of optima), this work gives us some cause for optimism.

4.4 Revisiting Algorithms

However, quite apart from the question as to whether NFL applies to 'real' problems, there remains a difficulty with relating NFL arguments to 'real' algorithms. These arguments tend to require that the search does not revisit previously seen points, which is a property that does not characterize real algorithms. Wolpert and Macready extend their result to allow for a search algorithm A that does revisit by defining a new algorithm A' that 'skips over' such points. The new algorithm can be described as a 'compacted' version of the original. The NFL theorem then clearly applies to the compacted version A', and can also be applied to A as long as the value m is re-interpreted as the number of distinct points seen. The table $\vec{c}$ must also be interpreted as relating only to the distinct points and not to possible repeats.

They then argue that the

> *real-world cost in using an algorithm is usually set by the number of* distinct *evaluations of f.* [314, emphasis added]

This is debatable, to say the least. In the real world, search algorithms do tend to revisit previously seen points, and such revisiting is *not* costless. The whole idea of tabu search [91], for instance, is predicated on the idea that revisiting is a bad idea, and that steps should be taken to limit it. Either we have to accept the fact of revisiting, and incur the expense of evaluating f for a point $\mathbf{x}$ that we have already seen (and evaluated) before, or we have to implement some mechanism for detecting that a revisit has occurred in order to save ourselves an unnecessary evaluation. The latter approach becomes more and more computationally demanding as the number of points visited increases.

The problem of revisiting is particularly acute in some implementations of a genetic algorithm. It was for this reason that Booker [26] and Davis [47] advocated diversity-inducing strategies—'reduced surrogate' and 'no-duplicate' policies respectively, as highlighted above in Section 2.3.8. Neither of these strategies guarantees that the next point is distinct from *all* those seen before, but they are relatively inexpensive ways of reducing the chance of revisiting, and both are claimed to enhance the performance of a GA.

4.4. REVISITING ALGORITHMS

That this is important seems intuitively sensible, but it also follows as a consequence of NFL. It is a straightforward argument from the NFL theorem, that if we have a decision between two algorithms, we should choose that algorithm which consistently revisits fewer points, since on average it cannot be worse. Of course, whether such an algorithm (i.e., a *consistent* one) exists is an open question, but GAs can generate an alarming number of revisits. One could make out a plausible *prima facie* case for the superiority of tabu search, for example, on the grounds of its revisiting performance.

It is also true that uniform random search is a simple approach with (for large search spaces) a very small probability of revisiting previous points until a substantial fraction of the space has been seen. Once again, therefore, in preferring GAs (or other heuristics) and ignoring random search, we make an implicit assumption as to the match between the problem and the method used.

4.4.1 Further reflections

Anecdotal evidence and personal experience suggest that *in practice*, random search has not been found to be as efficient and effective as heuristic methods that impose neighbourhood relations on the points in the search space. The results of Igel and Toussaint [132] now provide some support for the non-appliance of NFL arguments to 'real' problems. However, it is also true that few reported applications of GAs or other techniques such as NS bother to compare their results with those generated by random search. English's argument [70] suggests that we should consider this option much more seriously, not least because its routine use might help to identify properties of problems for which heuristics are superior.

It is also probable that some problem classes may be more easily solved by one technique than by another, although the extent to which this offers more than an 'appetizer' may be limited. As an example of this, Reeves [215] has shown that a particular class of deceptive functions that are hard for a GA can be optimized easily by a fairly general implementation of tabu search.

A related question, in the light of GA controversies, is the usefulness (or otherwise) of recombination. Fogel and Atmar [78] have argued that mutation is superior, and that crossover is of minor importance—in direct contradiction of Holland's original hypothesis! Eshelman and Schaffer [74] also considered this question, delineating what they believed to be 'crossover's niche' (roughly speaking, problems that are deceptive, but not too much) using schema-processing arguments. Höhn and Reeves [122, 123] have also put forward an argument based on a landscape perspective, which will be considered in Chapter 9.

NFL also has ramifications for traditional ways of evaluating heuristic performance. A popular way of comparing different search methods is to devise some test problem instances (often generated by some random process), and then to observe and analyse their performance on these test problems. This 'beauty contest' approach has already been critiqued independently by Hooker [128], but it is a practice that is still followed widely. In the light of the NFL theorem, it is a procedure that is clearly in need of revision—if the test instances are biased in some way, we will end up tuning 'our' heuristic to a particular set of benchmarks, only to find subsequently that it performs much less well in the more general case!

It would also seem reasonable that revisiting should be discouraged in general, so methods (such as tabu search) that allow this to be implemented simply and effectively would seem to be favoured. However, there are other considerations that the NFL theorem ignores—for example, the question of the computational complexity of an algorithm, and any computational overheads required by diversification strategies.

4.5 Conclusions

Opinion differs as to the extent of the significance of the NFL theorem. Some authors accord it virtually no practical significance at all, holding that while it applies on the average, 'real' functions have structure, and structure means that heuristics such as a GA have a leverage that makes them superior to random search. However, against that there are English's more recent developments of NFL to reckon with—arguments that make the usefulness of random search appear to be rather more considerable than is customarily assumed. The work of Droste *et al.*[64] also suggests that the leverage may be limited in both scope and effect.

Furthermore, while the existence of structure is undeniable for many practical problems, it is not so easy to pin down exactly what this means, nor how it might be possible to make use of it. Concurrently with the development of NFL-type arguments, there have been several attempts within the GA community to address this problem, and Chapters 8 and 9 will discuss some of this research.

4.6 Bibliographic Notes

David Wolpert's and Bill Macready's paper [314] is seminal for a study of this area. The archives of the GA-List (http://www.aic.nrl.navy.mil/galist/, predominantly in volume 9) will be helpful in clarifying some common misunderstandings concerning the implications of NFL. However, many readers may still find Joe Culberson's later paper [42] an easier route into understanding NFL ideas.

Nick Radcliffe and Patrick Surry originally raised the issues relating to representation and coding in [206], while Darrell Whitley and his colleagues have made several important contributions to this area [207, 310, 311]. Tom English is almost solely responsible for making the intriguing connections with information theory [68, 69, 70].

Chapter 5

GAs as Markov Processes

5.1 Introduction

A process or algorithm that has a random element is called a *stochastic process*. We can think of such a process as a sequence of random events occurring in time:

$$X_0, X_1, X_2, \ldots$$

Each of the $\{X_t\}$ is a random variable. The possible values that these variables can take are called the *states* of the system. For example, if we consider a sequence derived from rolling a single die repeatedly, the set of possible states is $\{1, 2, 3, 4, 5, 6\}$. On any one roll, we shall obtain one of these values. The probability that we obtain a particular value at a particular time is $1/6$. Each roll is independent of every other roll, but in many stochastic processes what happens at one time step is influenced by what has happened previously. This is certainly true for a genetic algorithm, in which the population at one time step clearly depends on what the population was at the previous time step. This means that at a given time t there is a probability distribution over the set of states, and that this distribution is a function of what states have occurred in previous time steps.

The simplest kind of dependence within a stochastic process is when the distribution for time t depends only on what happened at time $t-1$. If this is the case, then we have a *Markov process* and the sequence $X_0, X_1, X_2, \ldots$ forms a *Markov chain*. Let us consider the case where we have a finite set of s possible states, $\mathrm{E} = \{0, 1, 2, \ldots, s-1\}$. Suppose that at time $t-1$ state k occurred. Then there is some probability distribution for X_t, and this distribution might be different for each possible value of k. We can write out each such distribution like this:

$$\Pr[0|k], \Pr[1|k], \Pr[2|k], \ldots, \Pr[s-1|k]$$

where $\Pr[i|k]$ means the probability that state i occurs, given that state k occurred at the previous time step. This means that we can completely describe the system by s of these probability distributions, one for each k. We shall write each of these distributions as columns in an $s \times s$ matrix $\boldsymbol{Q}$. That is, the i,jth entry of $\boldsymbol{Q}$ is

$$Q_{i,j} = \Pr[i|j]$$

Notice that we can also represent particular states using vectors. State k corresponds to the vector $\boldsymbol{e}_k$, which contains a 1 in position k and zeros elsewhere. If the process is in state k at time $t-1$, then the probability distribution for time t is $\boldsymbol{Qe}_k$. More generally, if we know the probability of each state occurring at time $t-1$ then we can write this as a vector $\boldsymbol{p}(t-1)$, whose kth element is the probability of having state k. The probability distribution for time t is then simply

$$\boldsymbol{p}(t) = \boldsymbol{Q}\boldsymbol{p}(t-1)$$

If we start with some initial distribution $\boldsymbol{p}(0)$ over the set of states, E, then it follows that, for all t

$$\boldsymbol{p}(t) = \boldsymbol{Q}^t \boldsymbol{p}(0)$$

a version of the *Chapman-Kolmogorov* equation.[1] The matrix $\boldsymbol{Q}$ is called the *transition* matrix for the Markov chain, as it contains all the information about the probabilities of the different possible state transitions. A transition matrix is an example of a *stochastic* matrix: one in which the entries are all non-negative, and the columns sum to one.

Before we start to analyse genetic algorithms using the Markov chain framework, let us consider a couple of simple examples, to help familiarise ourselves with the concepts. We shall look at two simple optimization algorithms. The first algorithm starts with some initial random point in the search space, which it evaluates. This is our *current* point. We then generate another point, maybe a neighbour of the current one (if the search space has some topological structure). If this point is better than the current one

---

[1] It should be noted that many authors define $Q_{i,j} = \Pr[j|i]$, with $\boldsymbol{p}$ as a row vector, giving $\boldsymbol{p}(t) = \boldsymbol{p}(0)\boldsymbol{Q}^t$. That is, matrix multiplication takes place from the right. We have chosen our definition to remain consistent with material in later chapters, which will concern operators acting on (column) vectors.

5.1. INTRODUCTION

then it replaces it and becomes our new current point. If it is worse, however, then we have to decide whether we should risk accepting it. We adopt the following rules:

- If, last time we had a worse point, we accepted it, then only accept this time with probability 0.1.

- If, however, we didn't accept the last worse point we found, then accept this time with probability 0.5.

The probability values are parameters of the algorithm. Notice that the probability of accepting a worse point depends on whether or not we accepted the last worse point to come along. This tells us that we have a Markov process. What are the states of the system? These describe the behaviour of the algorithm, each time there is a probabilistic choice. We therefore have two possible states: *accept* and *reject*, indicating what happens when a worse point is generated. We are not including what happens when a better point is generated: this is automatic and not stochastic. The transition matrix for the system is:

$$Q = \begin{bmatrix} 0.1 & 0.5 \\ 0.9 & 0.5 \end{bmatrix}$$

where the first column (and row) corresponds to accepting, and the second column (and row) to rejecting. By simple observation of the matrix Q, we can deduce that each of the two states is *accessible*. That is, there is a non-zero probability that each state will eventually be reached. A transition matrix Q is called *primitive* if, for some t, the matrix Q^t contains all non-zero entries.

5.1.1 Steady-state behaviour

One of the questions we might want to investigate is: over a long period of time, how many worse points will be accepted, and how many rejected? We know from the Chapman-Kolmogorov equation that we can find the probability distribution over the states at time t by

$$p(t) = Q^t p(0)$$

and so we are interested in what happens as $t \to \infty$. There are some obvious worries. First, the equation depends on $p(0)$. It seems likely that for different choices of this vector, we shall obtain different long-term behaviours.

Second, how do we know that the limit exists? Maybe the system falls into a periodic behaviour pattern. Fortunately, neither of these worries need bother us, owing to the following remarkable theorem.

Theorem 5.1 (Perron-Frobenius) *Let Q be any primitive stochastic matrix. Then:*

1. $Q^\infty = \lim_{t \to \infty} Q^t$ *exists.*

2. *The columns of Q^∞ are identical.*

3. *Each column is the same as the unique probability vector q which satisfies $Qq = q$. That is, q is an eigenvector of Q with eigenvalue 1.*

4. $q = \lim_{t \to \infty} Q^t p(0)$, *regardless of the choice of $p(0)$.*

5. *Each entry of q (and therefore Q^∞) is strictly positive.*

A few calculations with the matrix Q describing our first algorithm gives

$$Q^6 = \begin{bmatrix} 0.3598 & 0.3557 \\ 0.6402 & 0.6443 \end{bmatrix}$$

which shows that the columns are already becoming similar after just six iterations. The theorem tells us that we can calculate Q^∞ directly by solving

$$\begin{bmatrix} 0.1 & 0.5 \\ 0.9 & 0.5 \end{bmatrix} \begin{bmatrix} x \\ 1-x \end{bmatrix} = \begin{bmatrix} x \\ 1-x \end{bmatrix}$$

which gives us

$$Q^\infty = \frac{1}{14} \begin{bmatrix} 5 & 5 \\ 9 & 9 \end{bmatrix}$$

and so $\lim_{t \to \infty} Q^t p(0) = (5/14, 9/14)$, independent of the initial distribution $p(0)$.[2] We therefore conclude that the ratio of acceptance to rejection in our algorithm (in the long-term) is $5 : 9$.

---

[2]Strict accuracy would require that we write $(5/14, 9/14)$ explicitly as a column vector. But here, as elsewhere, we shall refuse to be pedantic about this, believing that the reader is well able to interpret the notation correctly.

5.1. INTRODUCTION

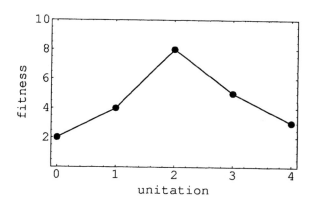

Figure 5.1: Function of unitation for hill-climbing algorithm.

5.1.2 Absorbing states

We shall now consider our second example, which is a hill-climbing algorithm. Our search space is the set of binary strings of length 4. The fitness function is a function of unitation (i.e., it depends only on the number of ones in the string—see Appendix A for further details) and is illustrated in Figure 5.1. As with our first example, the algorithm starts with a random point, and considers alternative points as replacements for the current one. In this algorithm we generate new points by mutating a single bit value in the current point. That is, moves take place between neighbours of Hamming distance 1. Each of the four bits is equally likely to be mutated. The new point is accepted if its value is greater than the value of the current point, otherwise it is rejected. From the graph of the fitness function, it seems clear that the hill-climber should find the optimum. Now we can analyse this using Markov chains.

In this algorithm, the random element comes in when we are generating our new point. Each of the Hamming-one neighbours of the current point is equally likely. The probability of generating a particular point in the search space depends only on what the current point is. We can therefore model our algorithm as a Markov process, having as states the different elements of the search space. This would give us $2^4 = 16$ states. In fact, because the fitness function is a function of unitation, we can simply consider the states to be the different unitation classes. Our set of states is therefore $\{0, 1, 2, 3, 4\}$. Consider what happens when the algorithm is in state 1. That means that the current string contains a single 1. Points generated from this string either have 2 ones, or none. The probability of obtaining 2 ones is 3/4. If this happens, then the new string will definitely be chosen, and we move to state 2. On the other hand, there is a probability of 1/4 that the string of all zeros will be generated. This won't be accepted, as it is worse than the current string. In this case, we stay in state 1. Calculating the possibilities for each state gives us the following transition matrix:

	0	1	2	3	4
0	0	0	0	0	0
1	1	1/4	0	0	0
2	0	3/4	1	3/4	0
3	0	0	0	1/4	1
4	0	0	0	0	0

There are a couple of things to notice about this matrix. Firstly, it is clear that the states 0 and 4 cannot be reached from any other states (the probabilities are zero in the respective rows). The only way that the algorithm might enter these states is if it happened to start in one of them. It would immediately move to another state at the next time step and never return. This means that the transition matrix is not primitive, and so the Perron-Frobenius theorem does not apply. Secondly, there is a probability of 1 that, once the algorithm has reached state 2 (the top of the hill) it will stay there. Any state with this property in a Markov chain is called an *absorbing* state. Absorbing states correspond to having a 1 on the main diagonal of the transition matrix. As long as there is some set of transitions from each state to an absorbing one, then it is clear that the long-term behaviour of the Markov chain is to arrive at one of the absorbing states and then stay there. For this situation, we have the following theorem, concerning the long-term behaviour.

Theorem 5.2 *Suppose the states of a Markov chain can be reordered such that the transition matrix has the following form:*

$$Q = \begin{bmatrix} I & R \\ 0 & S \end{bmatrix}$$

Then Q is called reducible, *and*

$$\lim_{t \to \infty} Q^t = \begin{bmatrix} I & (I - S^T)^{-1} R^T \\ 0 & 0 \end{bmatrix}$$

(where ^{T} indicates the transpose of a matrix).

5.1. INTRODUCTION

The matrix $(I-S^T)^{-1}$ is often called the *fundamental matrix* of the Markov chain, and it has some useful properties. Returning to our example, suppose we rearrange the states into the order $2, 0, 1, 3, 4$, to give transition matrix:

	2	0	1	3	4
2	1	0	3/4	3/4	0
0	0	0	0	0	0
1	0	1	1/4	0	0
3	0	0	0	1/4	1
4	0	0	0	0	0

Since there is only one absorbing state, the identity matrix in the top left-hand corner of the matrix is 1×1. S is a 4×4 matrix and R is 1×4. Applying the theorem gives:

$$\lim_{t \to \infty} Q^t = \begin{bmatrix} 1 & 1 & 1 & 1 & 1 \\ 0 & 0 & 0 & 0 & 0 \\ 0 & 0 & 0 & 0 & 0 \\ 0 & 0 & 0 & 0 & 0 \\ 0 & 0 & 0 & 0 & 0 \end{bmatrix}$$

which shows straight away that the long term behaviour (regardless of the initial state) will be to stay in state 2, the top of the hill. Actually, we can say more using this model. The fundamental matrix provides us with the expected number of time steps the process will take to reach the optimum.

Theorem 5.3 *Given a reducible transition matrix as in the previous theorem, let the expected time to absorption from a non-absorbing state k be denoted a_k. Let a be the vector containing these times, ordered according to the ordering of states in Q. Then*

$$a = (I - S^T)^{-1} 1$$

where 1 is the vector containing all ones.

Applying this theorem to our algorithm gives:

$$a = (2.3333, 1.3333, 1.3333, 2.3333).$$

Thus, the expected time to reach the optimum from states 0 and 4 is 2.3333, and the expected time from states 1 and 3 is 1.3333.

The main issues when analysing a Markov chain are therefore:

1. Are there any absorbing states?

2. If so, how long does it take to reach them?

3. If not, is the transition matrix primitive (and if so, what is its limiting distribution)?

We shall now apply these ideas to the study of genetic algorithms.

5.2 Limiting Distribution of the Simple GA

The first step in analysing a genetic algorithm as a stochastic process is to define the set of possible states that the algorithm can be in. We then need to confirm that the random sequence of states generated by a run of the algorithm does indeed form a Markov chain. That is, we need to check that the probability that the GA is in a particular state at a particular time depends at most on the state at the previous time step.

The natural candidate for describing the state of a GA is the population. At each generation we have a particular population, and the contents of that population depend (stochastically) on the previous generation. With this choice of state, we do indeed have a Markov chain. If the size of the search space is $|\mathcal{X}| = n$, and the population contains N individuals from the search space (possibly with duplications—remember that in general it is a multiset), then the number of possible states is:

$$\binom{n+N-1}{N}.$$

(This can be proven by induction, or by a combinatorial argument [182]). To keep things simple, we identify the search space $\mathcal{X}$ with the set of integers $\{0, 1, 2, \ldots, n-1\}$. Then we can represent populations as vectors:

$$\boldsymbol{v} = (v_0, v_1, \ldots v_{n-1})$$

in which v_k is the number of copies of individual $k \in \mathcal{X}$ in the population. Clearly

$$\sum_{k=0}^{n-1} v_k = N.$$

The states are all the possible non-negative integer vectors with this property. As we have remarked in Chapter 2, the fitness function can be thought of as a function $f : \mathcal{X} \to \mathbb{R}^+$.

5.2. LIMITING DISTRIBUTION OF THE SIMPLE GA

5.2.1 Effect of selection

Having defined the states of the GA Markov chain, we can calculate the transition matrix. It is easiest if we break this down into three stages—selection, mutation and crossover—so first we consider a GA that only employs selection. Such an algorithm introduces no new search space elements into the population. What would we expect to happen in this situation? Assuming that selection always tends to preserve the better elements of the population, and throw out the worse, we would expect that eventually the population would contain only copies of the best individual in the initial population. We can easily formalise this using the Markov chain framework.

Suppose our current population is v and we apply selection. We need to calculate the probability distribution over all possible next populations, as a function of v to give us our transition matrix. If we use fitness-proportional selection, the probability of any individual $i \in \mathcal{X}$ being selected is

$$\mathbf{Pr}[i|v]_1 = \frac{v_i f(i)}{\sum_{j \in \mathcal{X}} v_j f(j)}$$

where the subscript 1 indicates that we have a one-operator GA. To obtain the next generation, we take N samples of this distribution over $\mathcal{X}$. This is called a *multinomial* distribution. The probability of generating a population u in this way is

$$\mathbf{Pr}[u|v] = N! \prod_{i \in \mathcal{X}} \frac{\mathbf{Pr}[i|v]_1^{u_i}}{u_i!}$$

and this formula gives us our selection-only transition matrix. We want to know if this Markov chain has any absorbing states. Our guess was that *uniform* populations (those containing N copies of a single individual) would be absorbing. To check this, suppose that v is a population containing only copies of $k \in \mathcal{X}$. That means $v_k = N$ and $v_i = 0$ for all $i \neq k$. Then with fitness-proportional selection, we have

$$\mathbf{Pr}[i|v]_1 = [i = k]$$

and therefore

$$\mathbf{Pr}[v|v] = 1.$$

(Note again the use of the indicator function notation $[expr]$ as introduced in Chapter 2.)

Now, recall that having a 1 on the diagonal of the transition matrix corresponds to an absorbing state. This proves that all uniform populations are absorbing states of the selection-only genetic algorithm. Notice that the same is true for *any* selection method for which $\mathbf{Pr}[i|v] = [i = k]$ (i.e., where population v contains only copies of individual k). Since no selection method introduces new individuals into the population, then the conclusion holds.

5.2.2 Effect of mutation

Mutation involves changing one element of the search space to another, with a certain probability. Denote the probability that $j \in \mathcal{X}$ mutates to i by $U_{i,j}$. This gives us a $n \times n$ matrix U. The probability of generating an individual $i \in \mathcal{X}$ after selection, then mutation is

$$\mathbf{Pr}[i|v]_2 = \sum_{j \in \mathcal{X}} U_{i,j} \mathbf{Pr}[j|v]_1$$

and therefore the two-operator transition matrix is given by

$$\mathbf{Pr}[u|v] = N! \prod_{i \in \mathcal{X}} \frac{\mathbf{Pr}[i|v]_2^{u_i}}{u_i!}$$

Now suppose mutation is defined in such a way that $U_{i,j} > 0$ for all $i, j \in \mathcal{X}$. This would be the case, for example, with bitwise mutation acting on fixed-length binary strings. Then it is easy to see that $\mathbf{Pr}[i|v]_2 > 0$ for all $i \in \mathcal{X}$, and for any population v. That is, given any population v, there is a non-zero probability that the next generation will contain $i \in \mathcal{X}$, for any individual i. From this observation, we can conclude that the two-operator transition matrix contains no zero terms—and therefore contains no absorbing states. The matrix is primitive, so that (applying Theorem 5.1) there is a limiting distribution over the set of possible populations and, in the long-term, there is a non-zero probability that any given population might arise. The GA with selection and mutation does not 'converge'. It will wander through every possible population. However, some populations may be much more likely to occur than others, and we shall consider this possibility later.

5.2.3 Effect of crossover

The addition of a crossover operator will not change this conclusion. As long as mutation can generate any individual from any population, the transition

5.2. LIMITING DISTRIBUTION OF THE SIMPLE GA

matrix will be primitive. Suppose we denote the probability that i and j cross to form k by $r(i, j, k)$. Then the probability of generating individual k from population v in the three-operator GA is:

$$\mathbf{Pr}[k|v]_3 = \sum_{i,j \in \mathcal{X}} r(i,j,k) \mathbf{Pr}[i|v]_2 \mathbf{Pr}[j|v]_2$$

This probability will be non-zero provided that, for any $k \in \mathcal{X}$ there exists at least one pair i, j such that $r(i, j, k) > 0$. For most crossovers this will be true, since at least $r(k, k, k) > 0$ for each k. The transition matrix is now given by:

$$\mathbf{Pr}[u|v] = N! \prod_{i \in \mathcal{X}} \frac{\mathbf{Pr}[i|v]_3^{u_i}}{u_i!}$$

which is therefore primitive. Note that, if we assume crossover takes place before mutation instead of afterwards, it is easy to verify that we still obtain a primitive transition matrix.

When we have mutation in a GA (with or without crossover), we know that the long-term behaviour of the algorithm is described by a limiting distribution over the possible states. This distribution is a vector q, in which q_v is the probability associated with population v. The following theorem gives us a formula for q.

Theorem 5.4 (Davis-Principe [50]) *Suppose Q is the primitive transition matrix for a genetic algorithm with non-zero mutation. Then the limiting distribution is given by*

$$q_v = \frac{|Q_v - I|}{\sum_u |Q_u - I|}$$

where the matrix Q_u is derived from Q by replacing the uth column of Q with zeros.

5.2.4 An example

We can illustrate this result by a simple example. Suppose our search space is $\{0\,0, 0\,1, 1\,0, 1\,1\}$ which we identify with $\mathcal{X} = \{0, 1, 2, 3\}$ by considering each string to be a binary number. The fitness function is:

$$\begin{aligned} f(0\,0) &= 1 \\ f(0\,1) &= 2 \\ f(1\,0) &= 3 \\ f(1\,1) &= 4 \end{aligned}$$

Mutation is performed bitwise with probability 0.1 per bit. This gives us a mutation matrix

$$U = \begin{bmatrix} 0.81 & 0.09 & 0.09 & 0.01 \\ 0.09 & 0.81 & 0.01 & 0.09 \\ 0.09 & 0.01 & 0.81 & 0.09 \\ 0.01 & 0.09 & 0.09 & 0.81 \end{bmatrix}$$

To keep things easy to calculate, consider a population of size $N = 4$. This means there are $\binom{7}{4} = 35$ possible populations. It is straightforward to enumerate these and calculate the transition matrix for the GA with proportional selection and mutation. Using the determinant formula for the limiting distributions, we can calculate that the populations with the highest probabilities are:

population	probability
$(0, 1, 0, 3)$	0.102
$(0, 0, 1, 3)$	0.124
$(0, 0, 0, 4)$	0.143

We can see that the most probable case is the uniform population containing the optimum. However, this population occurs only 14.3% of the time (that is, about one in seven). There are 15 possible populations which do not contain the optimum at all. Summing the probabilities for these states tells us that the long-term behaviour of the GA has non-optimal populations 17% of the time. We cannot expect a run of this GA to converge to a fixed population, then, no matter how long we run it. This is shown in practice in Figure 5.2 which shows a typical run of this genetic algorithm. The actual proportion of uniform optimal populations (containing only the optimum) and suboptimal populations (containing no copies of the optimum) are shown over 20000 generations, together with the theoretical limiting values for these states (14.3% and 17% respectively).

5.3 Elitism and Convergence

One way in which we could easily make a genetic algorithm converge is to keep a record at each time step of the best individual seen so far. Since this sequence of values never decreases and since it is bounded above by the value of the optimum, it must converge. This doesn't actually buy us very much, however, because exactly the same modification can be made to any search algorithm, including random search. It would be nicer if we could build something into the GA that would naturally ensure convergence.

5.3. ELITISM AND CONVERGENCE

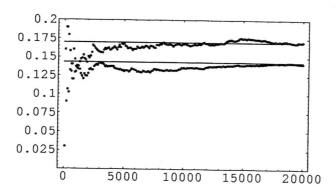

Figure 5.2: Proportion of uniform optimal populations (containing only the optimum) and suboptimal populations (containing no copies of the optimum) for the example GA, shown over 20000 generations, together with the theoretical limit values for these states (14.3% and 17% respectively).

This is what *élitism* does. Any genetic algorithm is said to be élitist if the following condition holds for all times t:
Let k be the best individual at time t. Then at time $t+1$ either:

1. k is in the population, or

2. something better than k is in the population

An example of a genetic algorithm that is naturally élitist is a steady-state GA in which the new individual always replaces the current worst individual. Since the current best individual always remains, the algorithm is élitist.

Any genetic algorithm can easily be made élitist by the following adaptation:

1. record the current best individual k in the population

2. generate the next population as usual

3. if there is nothing in the new population better than k, add k to the new population, replacing some other individual.

The idea is simple: the value of the best individual in the population is a non-decreasing sequence, bounded above. Therefore this sequence must converge. However, when we talk about sequences of random variables, we must be more precise about what we mean by 'convergence'. There is always a chance that the random sequence might not behave as expected. For example, consider the sequence of numbers generated by repeatedly rolling a die and recording the highest number found so far at each roll. Again we have a sequence of numbers that is non-decreasing and bounded above, and we suspect that the sequence will always converge to 6. However, since this is a random sequence, there is the possibility that a six is never rolled at all! In what sense then, can we say that such a series converges? There are a number of possible related definitions. The one we shall take here is that a sequence of random variables $X_0, X_1, X_2, \ldots$ converges *almost surely* to the random variable X if

$$\mathbf{Pr}[\lim_{t \to \infty} |X_t - X| = 0] = 1$$

That is, the sequence converges to a limit, if the probability is one that it is the same as the limit as $t \to \infty$.

One way in which we can prove that a real-valued sequence converges almost surely, is to show that it forms a *supermartingale*. First, suppose that we have an underlying stochastic sequence, $X_0, X_1, X_2, \ldots$. Now let D map the states of this sequence to real numbers. We shall use the notation $D_t = D(X_t)$. Then the sequence $D_0, D_1, D_2, \ldots$ is a real-valued stochastic sequence. If the function D is bounded, and we have the property

$$E[D_{t+1}|X_t] \le D_t$$

then the sequence is a supermartingale. Non-negative supermartingales have the property that they converge almost surely to some limit.

We can now prove something about élitist genetic algorithms:

Theorem 5.5 (Rudolph [247]) *Let X_t be the population at time t, and let $f(X_t)$ be the fitness of the best solution in X_t. If a genetic algorithm has mutation probability $0 < \mu < 1$, arbitrary crossover and selection, but is élitist, then the sequence*

$$D_t = f^* - f(X_t)$$

(where f^* is the optimal solution) is a non-negative supermartingale which converges almost surely to zero.

The proof of this theorem is fairly straightforward. The sequence we are considering at each time step is the difference between the optimal solution and the current best. To see that the sequence is a supermartingale, we need to realise that élitism guarantees that the 'best in the population'

5.3. ELITISM AND CONVERGENCE

sequence is non-decreasing, so that the sequence converges almost surely to some limit. To show that the limit is zero, however, needs something a bit stronger. As long as mutation is greater than zero and less than 1, there is a non-zero probability that some member of the population mutates to become the optimal solution. This means that the *expected* distance of the current best from the optimum is a strictly decreasing sequence, i.e.,

$$E[D_{t+1}|X_t] < D_t.$$

The value of each D_t is non-negative, and thus the sequence tends to zero. This proves that a genetic algorithm with élitism will converge (almost surely) to a population containing the optimum.

Consider the following example. Suppose we run an élitist GA on the *Onemax* problem (one where fitness is the number of 1s in the string—see Appendix A). The genetic algorithm uses stochastic universal sampling (SUS), mutation is performed on exactly one bit per string and there is no crossover. Elitism is implemented by inserting the best of the previous generation in place of the worst in the current population. We know from Theorem 5.5 that the GA will converge (almost surely) to a population containing the optimum, and we can illustrate this by deriving an upper bound on the expected time for this to happen. The technique we shall use is to construct a Markov chain that converges to the same limit, but more slowly than the GA. We choose this Markov chain to be easier to analyse than the original, but without its being too inaccurate.

Firstly, we note that the form of mutation we use means that improvements from one generation to the next can only come about through mutating one of the current best individuals. One of the useful properties of SUS is that it is guaranteed to select such an individual. At worst, there is only one of these. So if we assume that this worst case always holds, we shall obtain an upper bound. However, this case is equivalent to having a single individual hill-climbing, just as in the earlier example (although there we had a different fitness function), since élitism means that we can't do worse than our current best. We can therefore write down the transition probabilities as:

$$\mathbf{Pr}[i|j] = \begin{cases} \frac{i}{\ell} & \text{if } i = j \\ 1 - \frac{i}{\ell} & \text{if } i = j + 1 \\ 0 & \text{otherwise} \end{cases}$$

for our simplified Markov chain. The only absorbing state is, of course, the optimal solution. Using the same method as for the hill-climbing algorithm, we can calculate the expected time to absorption, for different

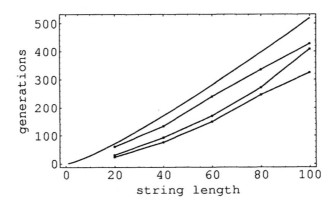

Figure 5.3: Average number of generations until an élitist GA finds the optimum in the *Onemax* problem. The upper line is the theoretical upper bound. Population sizes 1, 25 and 100 are shown (population size 1 is the slowest, 100 is the quickest). The data were averaged over 30 runs.

starting states, by applying Theorem 5.3. To make sure we do have an upper bound, we can use the expected time to absorption starting from state **0** (the worst case).

Given the way we constructed our simple Markov chain, this bound tells us nothing about the advantages of having a population, and the effects of the population size. These will come from the fact that at any time we might have several strings of maximal fitness within a population. Therefore, any one of these has the potential to improve. Figure 5.3 illustrates the difference a population can make. The upper line gives the theoretical upper bound, calculated from the simplified Markov chain. Runs for populations of size 1, 25, and 100 are shown (averaged over 30 runs each), for different values of ℓ. Note that a population of size 1 is very close to being the simplified Markov chain that was analysed. It runs slightly faster because it is starting with a random individual, not with the worst one. As we suspected, having a larger population increases further the chances of finding the optimum. It should be remembered, however, that the graph shows the average number of *generations* until the optimum was found. Of course, to calculate the number of fitness evaluations, we need to multiply by the population size.

5.4 Annealing the Mutation Rate

If a genetic algorithm has selection, no mutation, and possibly crossover, we know that it will eventually converge to a uniform population: one containing multiple copies of an individual.[3] Mutation is used to add some exploratory power, but with the drawback, as we have seen, that the algorithm no longer converges. It would seem to be a good idea, then, to reduce the mutation rate slowly as the algorithm runs. This would enable some exploration to be done in the early stages of the search, but (hopefully) allow the population to converge in the long term. If we reduced mutation sufficiently slowly, it seems possible that we might be able to guarantee a discovery of the optimum in the early stages, and then converge onto it in the later stages of the search. This idea has strong parallels with the simulated annealing algorithm, in which there is a *temperature* parameter that reduces over time. High temperatures correspond to large amounts of exploration, while low temperature forces convergence. It is known that, given a suitable 'cooling schedule', simulated annealing will converge on the global optimum [154]. Is it possible that the same thing will happen for a genetic algorithm?

Unfortunately, there are problems. The most obvious problem is that in the case of zero mutation, the absorbing states of the Markov chain are each of the possible uniform populations. So there is no guarantee that the population we reach will be at all optimal [50]. There is also the technical problem of how quickly the mutation rate should be reduced, even to achieve this limit. The whole analysis becomes more difficult, because we now have a situation where the probability of moving from one state to another depends upon time. This means that instead of having a single transition matrix, we have a whole sequence of them, one for each time step. This kind of Markov chain is called *inhomogeneous*. (The static situation, with a fixed mutation rate, is called *homogeneous*.)

For the inhomogeneous Markov chain, we can proceed as follows. Firstly, we consider the limiting distribution for the GA with fixed mutation, and work out the limit of that distribution as the mutation rate tends to zero. Secondly, we try to find a suitably slow cooling schedule for the mutation rate to ensure that this limit is achieved in the inhomogeneous case. The zero-mutation limit can be calculated, and (as indicated above) this results in non-zero probabilities for each uniform population. A suitable cooling

---

[3]Actually, it is just a conjecture that this is the case with crossover as well as selection. It is known to be true if fitness is a linear function.

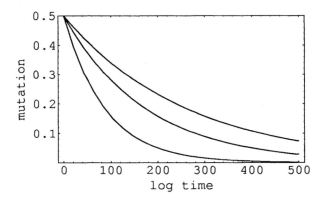

Figure 5.4: Cooling schedules for mutation, guaranteeing that the zero-mutation distribution over uniform population is achieved. Results for population sizes $N = 10, 20, 30$ are shown—the smaller the population size, the more rapid the cooling. Note that time is shown on a logarithmic scale (to base 10).

schedule that guarantees that the inhomogeneous GA converges to the zero-mutation limit is

$$\mu(t) \geq \frac{1}{2} t^{-\frac{1}{N\ell}}$$

where N is the population size [50]. These results are bad news for the idea of annealing. Firstly, annealing does not guarantee convergence to the optimum, only to some uniform population. Secondly, the cooling schedule is ridiculously slow for real sized problems. Figure 5.4 shows plots of the cooling schedule for $\ell = 20$, for various population sizes. Even with a population as small as 10, it takes 10^{400} generations to reach a mutation rate of 0.005. This is clearly impracticable! The alternative is to have a quicker cooling schedule, but then we run the risk that the optimal population may have zero probability of occurring. The reason for this is that, the more quickly mutation drops to zero, the more like the selection-only algorithm it becomes. And the selection-only algorithm converges to a uniform population containing some individual from the initial population, which almost certainly isn't the global optimum.

5.4.1 Further annealing

We could try to improve matters by simultaneously annealing the crossover rate and selection strategy. The probability of applying crossover could go to zero, and selection could become progressively stronger. By *strong* selection, we mean that there is a bias towards selecting the higher fitness individuals. The strongest selection of all chooses only those individuals that have the highest fitness in the population. If we arrange things so that selection strength increases up to this maximum, our algorithm becomes more and more élitist, and we might be able to force convergence on the optimum. It turns out that this can indeed be done, provided the population is suitably large. (Small populations can't carry enough diversity to allow the optimum to be found.) There is actually a critical population size (depending on the fitness function), above which this idea can be made to work. The population size needn't be unfeasibly large: $N \sim \mathcal{O}(\ell)$ will work for functions of bounded complexity. However, suitable cooling schedules are difficult to calculate, and the whole theory is based on a deep analysis of inhomogeneous Markov chains [34, 256, 257]. It is not known whether a practical algorithm can be built along these lines.

5.4.2 Adaptive mutation

A third strategy might be to have an adaptive mutation rate. That is, we try to organise things so that when the population is nearly uniform, we increase the mutation rate to try to force further exploration. Conversely, when there is a lot of diversity, we reduce the mutation rate to encourage exploitation of what has been found. However, in doing this, we have moved away from a Markov process. The value of mutation will depend not just on time, but on the whole history of populations that have occurred so far. We now have what is called a *random system with complete connections*. We shall not try to develop this theory here, but simply state some results for a particular self-adaptive scheme.

Suppose that, instead of having a single mutation rate, we have a separate mutation rate for each bit position. That is, we have a mutation vector

$$\mu = (\mu_0, \mu_1, \ldots, \mu_{\ell-1})$$

where μ_k is the probability of mutating bit k. We adjust this vector at each time step by calculating the amount of diversity at each bit position in the population. For some parameter $\epsilon > 0$, we say that bit position k is sufficiently diverse if the proportion of ones in position k is between $1/2 - \epsilon$ and $1/2 + \epsilon$. If this is the case, then we shall move the mutation rate μ_k towards some minimal level θ by setting the new mutation rate to

$$\mu'_k = \frac{\mu_k + \theta}{2}$$

If the bits in position k are not sufficiently diverse, we increase μ_k towards some maximum Θ by setting

$$\mu'_k = \frac{\mu_k + \Theta}{2}$$

where $\Theta > \theta$. Doing this for each bit position gives an update of the whole mutation vector [3].

It turns out that this form of self-adaptation, while it may (or may not) improve the searching abilities, does not alter the convergence properties of the algorithm (in general terms). That is, we do not have convergence to the optimum unless some form of élitism is introduced. The general argument follows the same lines as before: non-zero mutation ensures that it is possible to reach anywhere in the search space, and élitism guarantees that once the optimum is found, it is kept.

In fact, it is possible to relax élitism a little. Suppose the difference between the optimal fitness and the next best solution is λ. Then if we carry the best of one generation (at time t) forward to the next generation (at time $t+1$) only when

$$f(X_t) > f(X_{t+1}) - \lambda$$

we have a *threshold acceptance* algorithm [2]. Recall that $f(X_t)$ is the fitness of the best individual in population X_t. Given the choice of λ, it is obvious that once the optimum is discovered, it will always be carried forward. We shall still have convergence to a population containing the optimum.

5.5 Calculating with Markov Chains

One of the main problems with modelling genetic algorithms as Markov chains is that the matrices involved are so large that calculations for realistic systems are infeasible. We have already seen that the number of possible populations of size N, over a search space of n points is

$$\binom{N+n-1}{N}.$$

The following table illustrates how large this number becomes even for modest problems.

5.5. CALCULATING WITH MARKOV CHAINS

population size	search space	possible populations
10	2^{10}	10^{23}
20	2^{10}	10^{41}
20	2^{20}	10^{102}
50	2^{50}	10^{688}

To be able to perform calculations on specific Markov chains, we need some way of simplifying the matrices involved. One way to do this is to reduce the number of states, perhaps by grouping together states that are in some sense similar. We would then have a Markov chain with a transition matrix showing the probability of going from one group of states to another. It is impossible, in general, to do this completely accurately, since no two states will usually be exactly the same. Instead, we seek a method that will group together states that are approximately the same, with only a small loss of accuracy. Such a method was introduced by Spears and De Jong [268]. There are two questions that must be addressed:

1. How do we decide that two states are sufficiently similar to be grouped?

2. How to we calculate the transition probabilities between the new groups?

We shall consider the second question first. A suitable answer to the similarity problem will emerge from this.

Suppose that we have decided to put states u and v together to form a group. We need to calculate the following sets of probabilities:

1. The probability of going from some other state to the new group.

2. The probability of going from the new group to some other state.

3. The probability of staying within the new group.

We shall take these one at a time.

1. **The probability of going from another state to the new group.**
   Suppose we are in some state k. We wish to know the probability that we shall end up in a new group comprising states u and v. Clearly, we shall reach the new group if we go either to state u or to state v. Therefore
   $$\mathbf{Pr}[\{u,v\}|k] = \mathbf{Pr}[u|k] + \mathbf{Pr}[v|k]$$

   In other words, we can add rows u and v of our transition matrix together to give the row of our new reduced matrix.

2. The probability of going from the new group to another state.
Now suppose we are in the new group state. We want to find the probability that we move from this state to some state k. There are two possibilities. Either we are actually in state u and move to state k, or else we are in state v and go to k. But there's a problem here. In our new reduced Markov chain, we don't know which of u and v we are currently in. All we know is that we happen to be in one of these states. There would be no problem, of course, if the probability of moving from u to k was the same as moving from v to k, since we could just take the probability from either of these states. In other words, if columns u and v are identical in the transition matrix, then the column for our reduced matrix is the same as either one of them.

However, it is not very likely that there are many pairs of states with identical columns. (These would correspond to two distinct populations, for which the effects of the genetic algorithm are identical.) So here is where we introduce an approximation. What if we merge columns that are *nearly* identical? We can measure how similar two columns are by looking at the squared error, to see if it is less than some threshold, ϵ:

$$\|u - v\|^2 = \sum_k (\mathbf{Pr}[k|u] - \mathbf{Pr}[k|v])^2 < \epsilon$$

If this is the case, then we have candidate states for merging. Notice that this also answers our first question, of how we pick which states to merge.

We now have to consider what values to use for the probability of moving from our group of states to state k. When the columns are identical it is easy: just use one of the columns. Now we have columns which may be slightly different. One possibility is simply to average the two columns, but this leads to unacceptable inaccuracies. These arise because one of the two states in the group may be much more likely to occur than the other. If this is the case, the new probabilities should be weighted towards that state's column. How can we tell which state is more likely to occur? One way to obtain an estimate of this is to examine the *rows* of the transition matrix. Recall that row u of the matrix corresponds to the probabilities of entering state u. So if we add up the probabilities in row u and compare this total with the sum of the probabilities in row v, we have an estimate of which state is most likely to occur. Moreover, we can use these totals as weights

5.5. CALCULATING WITH MARKOV CHAINS

in a weighted average of the columns u and v. That is, if we calculate

$$w_u = \sum_j \Pr[u|j]$$
$$w_v = \sum_j \Pr[v|j],$$

we can set the new column in our reduced matrix to be

$$\Pr[k|\{u,v\}] = \frac{w_u \Pr[k|u] + w_v \Pr[k|v]}{w_u + w_v}$$

Note again that this is only an estimate of the exact probability, and it is the point where the reduced model becomes an approximation.

3. **The probability of staying within the new group.** The final value we need to calculate is the probability that, once we are in our new group state, we stay there. Once again, the problem arises that we don't actually know which of the two states we are in. Again, we solve this problem by estimating the probability of being in either state by the sums of the corresponding rows. Thus we set

$$\Pr[\{u,v\}|\{u,v\}] = \frac{w_u(\Pr[u|u] + \Pr[v|u]) + w_v(\Pr[u|v] + \Pr[v|v])}{w_u + w_v}$$

5.5.1 Some examples

We shall now illustrate this process with a simple example. Consider the following transition matrix, $Q =$

	0	1	2	3	4
0	0	0	0	0	0
1	1	1/8	0	0	0
2	0	3/4	7/8	3/4	0
3	0	0	0	1/4	1
4	0	1/8	1/8	0	0

We observe that states 1 and 2 have similar columns (the squared error is 1/32), so we shall merge these to form a new state, $\{1, 2\}$. States 0, 3 and 4 will remain unaffected by this process. Firstly, then, we shall consider the row of our reduced matrix corresponding to our new state. This row will contain the probabilities that the new state is reached from some other state. It is found by adding the entries of rows 1 and 2 together. That is:

$$\mathbf{Pr}[\{1,2\}|0] = 1$$
$$\mathbf{Pr}[\{1,2\}|3] = 3/4$$
$$\mathbf{Pr}[\{1,2\}|4] = 0$$

Secondly, we consider the column of the new matrix. To calculate this, we must first find the appropriate weights w_1 and w_2 by adding up the values in the corresponding rows.

$$w_1 = 1 + 1/8 + 0 + 0 + 0 = 9/8$$
$$w_2 = 0 + 3/4 + 7/8 + 3/4 + 0 = 19/8$$

Clearly, it is about twice as likely that state 2 will be entered than state 1. So we weight the column of our new matrix appropriately:

$$\mathbf{Pr}[0|\{1,2\}] = \frac{(9/8)0 + (19/8)0}{9/8 + 19/8} = 0$$
$$\mathbf{Pr}[3|\{1,2\}] = \frac{(9/8)0 + (19/8)0}{9/8 + 19/8} = 0$$
$$\mathbf{Pr}[4|\{1,2\}] = \frac{(9/8)(1/8) + (19/8)(1/8)}{9/8 + 19/8} = 1/8$$

Finally, we need to work out the estimate of the probability that the system will stay in state $\{1,2\}$.

$$\mathbf{Pr}[\{1,2\}|\{1,2\}] = \frac{(9/8)(1/8 + 3/4) + (19/8)(0 + 7/8)}{9/8 + 19/8} = 7/8$$

This gives us a reduced transition matrix $\mathbf{Q'} =$

	0	{1,2}	3	4
0	0	0	0	0
{1,2}	1	7/8	3/4	0
3	0	0	1/4	1
4	0	1/8	0	0

If we had an initial distribution vector $\mathbf{p} = (p_0, p_1, p_2, p_3, p_4)$ we can produce a reduced version

$$\mathbf{p'} = (p_0, p_1 + p_2, p_3, p_4)$$

5.5. CALCULATING WITH MARKOV CHAINS

We can then estimate $Q^t p$ for different times t using the reduced system $(Q')^t p'$.

Now we consider some more serious examples. Suppose we have a genetic algorithm with fitness-proportional selection and bitwise mutation. To construct the transition matrix, it is necessary to list all the possible populations of a given size for the given search space. The following algorithm will do this, assuming a population size of N, and a search space of n individuals [181].

```
The initial population is v = (N, 0, ..., 0).
Given a population v, update it as follows:
    1.  Let h = min{i|v_i ≠ 0}
    2.  Let τ = v_h
    3.  Let v_h = 0
    4.  Let v_1 = τ - 1
    5.  Let v_{h+1} = v_{h+1} + 1
If v_n == N then this is the final population.
```

It is then necessary to work out the probability that the genetic algorithm will change one population into another. We shall develop the details of this probability distribution in the next chapter, for genetic algorithms with and without crossover. We then use the state-merging idea to reduce this matrix. To illustrate what can be achieved, we can use a simple one-pass algorithm:

```
For i = 1 to maxstates do
    if state i has previously been merged then
        skip
    else
        Find state j > i, such that ||i - j||² < ε.
        If found, merge states i and j.
```

As a result of this algorithm, the states will be grouped into pairs. Each pair will have a squared error of less than the parameter ϵ. This means that the number of groups is at most one half of the original number of states. Obviously, the bigger ϵ is, the more groups we shall obtain, up to this maximum.

To start with, we consider a small system. The string length is 2, with fitness function

$$f(0\,0) = 1$$
$$f(0\,1) = 2$$

ϵ	states removed	t = 16	t = 64
0.00	0	0.8156	0.8297
0.01	0	0.8156	0.8297
0.02	3	0.8182	0.8326
0.03	9	0.7952	0.8037
0.04	10	0.7955	0.8043

Table 5.1: $n = 4, N = 4$. Probability of being in an optimal population after 16 and 64 generations, calculated with different reduced models. Fitness function is $f(0\,0) = 1, f(0\,1) = 2, f(1\,0) = 3, f(1\,1) = 4$.

$$f(1\,0) = 3$$
$$f(1\,1) = 4$$

The mutation rate is 0.1. We have studied this system before, with a population of size 4. We concluded that, in the long term, there is a probability of 17% that the population will not contain a copy of the optimum. To illustrate the effects of our one-pass algorithm on this problem, different values of ϵ were chosen, giving different degrees of state reduction. The probability of being in an optimal state (that is, in a population containing a copy of the optimum) after a certain number of generations was then calculated using the reduced system. It was assumed that the original population contained only copies of (0 0). When $\epsilon = 0$, we have no merging of states, and this case gives us the exact results. The approximations found with different values of ϵ are shown in Table 5.1.

The first thing to notice from the table of results is that, for $t = 64$, we have practically reached the long-term probability of 83% of having an optimal population. A value of $\epsilon = 0.02$ is required before we find any reductions in states. The value $\epsilon = 0.04$ gives a reduction of 10 out of the 35 states, while maintaining an accuracy of within 2%.

Accuracy improves as the population size increases. The reason for this is that in the increased population, stochastic effects are averaged out; the larger the population is, the more deterministic the system looks. We shall study why this happens in the next chapter. Results for this system with $N = 10$ are shown in Table 5.2. This time there are 286 possible populations. We can see that $\epsilon \geq 0.03$ reduces the number of states to nearly a half of the original, which is the best we can expect from a one-pass algorithm. Accuracy is now within 0.5%.

5.5. CALCULATING WITH MARKOV CHAINS

ϵ	states removed	t = 16	t = 64
0.00	0	0.9888	0.9908
0.01	137	0.9879	0.9893
0.02	139	0.9874	0.9887
0.03	142	0.9867	0.9879
0.04	142	0.9853	0.9862

Table 5.2: $n = 4, N = 10$. Probability of being in an optimal population after 16 and 64 generations, calculated with different reduced models. Notice the increase of accuracy associated with a (relatively) large population. Fitness function is $f(0\,0) = 1, f(0\,1) = 2, f(1\,0) = 3, f(1\,1) = 4$.

The two final examples are of two different problems with a string length of 3. The two functions are:

	f_1	f_2
0 0 0	1	18
0 0 1	2	12
0 1 0	3	15
0 1 1	4	4
1 0 0	5	12
1 0 1	6	5
1 1 0	7	8
1 1 1	8	20

The function f_2 has a local optimum (with respect to a bit-flip neighbourhood) at **0**. Recall that we are assuming the initial population to be made up of copies of just this string. We therefore guess (rightly) that the probability of reaching the optimum is fairly low for this fitness function, compared to the other examples. The results are given in Table 5.3. We again have $N = 4$, which is a relatively small population size. The number of possible populations is 330. With $\epsilon = 0.04$ we achieve nearly the maximum number of merged states, with an accuracy of around 2.5%.

We could reduce the number of states much further by grouping together more than just two states. To achieve this, we could use a multi-pass version of our merging algorithm:

```
Repeat until no states are merged
    Find states i and j, such that ||i − j||² < ε.
    If found, merge states i and j.
```

ε	states removed	t = 16	t = 64
0.00	0	0.5879	0.6075
0.01	112	0.5762	0.5913
0.02	147	0.5775	0.5923
0.03	158	0.5874	0.6002
0.04	160	0.5851	0.5966
ε	states removed	t = 16	t = 64
0.00	0	0.2260	0.3490
0.01	88	0.2280	0.3450
0.02	141	0.2373	0.3483
0.03	159	0.2343	0.3471
0.04	160	0.2329	0.3412

Table 5.3: $n = 8, N = 4$. Probability of being in an optimal population after 16 and 64 generations, calculated with different reduced models. Results for f_1 are shown above, f_2 below.

The multi-pass version can often eliminate the vast majority of states with relatively little loss of accuracy. However, it is worth mentioning that even this method will quickly become limited. When faced with 10^{688} possible states, there is very little one can do to make the system manageable. It is possible, though, to use this method to calculate the short-term behaviour of small systems with reasonable efficiency and accuracy.

5.6 Bibliographic Notes

The general tools of Markov chain analysis can be found in summarised form in Günter Rudolph's book *Convergence Properties of Evolutionary Algorithms* [247]. Any general textbook on stochastic processes (for example, [108]) would cover similar material.

Markov models for complete, non-trivial genetic algorithms, were developed independently by Thomas Davis and Jose Principe[50] and by Michael Vose (and co-workers)[182]. The Vose model looks at the detailed underlying equations of the system and has been substantially developed in recent years. This approach is the subject of the next chapter. The qualitative results presented in this chapter are chiefly due to Davis and Rudolph. Section 5.2 presents the main results of Davis and Principe [49], including the limiting distribution theorem. Section 5.3 covers some of Rudolph's results

5.6. BIBLIOGRAPHIC NOTES

on population based algorithms with finite search spaces. The principal result here is that a genetic algorithm typically requires élitism to converge (almost surely). The use of the supermartingale technique in analysing genetic algorithms is due to Rudolph. He also provided the proof that stochastic universal sampling is guaranteed to select the best individual in a population, which was used in the analysis of *Onemax*. This analysis should be compared to that of Mühlenbein[174], who analyses a hill-climbing algorithm (the '(1+1)-EA') in a similar fashion, although he takes it further, deriving an approximation for the expected number of function evaluations needed to reach the optimum. This was made more rigorous by Rudolph, and was illustrated experimentally by Thomas Bäck[10, 12]. The books by Rudolph and Bäck deal with a variety of evolutionary algorithms, including those based on individuals as well as on populations, and algorithms defined over continuous search spaces.

The idea of annealing the mutation rate (described in section 5.4) goes back to [50], which attempted to adapt the proof of convergence for simulated annealing to GAs. The cooling schedule presented here is derived in that paper, as is the proof that annealing the mutation rate cannot of itself guarantee convergence to the optimum. The conjecture in [50] was that annealing the selection strength and crossover rate as well as the mutation rate might do the trick, and an attempted proof of this was presented by Suzuki [276]. This has since been shown to be defective by Lothar Schmitt [256, 257], who proposed an alternative proof, and showed that there is a critical population size needed to achieve this result. This was independently proved by Raphael Cerf [34], who used a different technique based on perturbing the desired limiting system, and applying the Freidlin-Wentzell theory of stochastic perturbations. The analysis of the self-adaptive mutation scheme comes from Alexandru Agapie [3], who also proposed, and proved convergence for, threshold acceptance élitism [2].

Section 5.5 is based on the work of Bill Spears [268] who, with Ken De Jong, developed the one-pass and multi-pass 'lumping' algorithms, which have been presented here in simplified form. The results stated here have been produced independently, and are restricted to the one-pass algorithm applied to a GA with selection and mutation. Spears' work contains examples of the multi-pass algorithm and is applied to GAs with crossover, with various fitness functions. The algorithm for generating all possible populations in sequence comes from the useful book by Nijenhuis and Wilf[181].

Exercises

1. Suppose we have a modified random search algorithm: Let $a = 0$.
   Repeat:
   Generate a random point $x \in \mathcal{X}$.
   If $f(x) > a$ then let $a = f(x)$.

   Show that the value of the variable a converges almost surely to the optimal value of f.

2. Show that the stochastic universal sampling method of selection is guaranteed always to select the optimal member of a population.

3. Let $\mathcal{X}$ be the set of binary strings of length ℓ. A fitness function is *linear* if it can be written as
   $$f(x) = \sum_i a_i x_i$$
   for some values $a_1, \ldots, a_\ell$. Show that applying uniform crossover to a population of strings does not affect the average fitness of the population if the fitness function is linear.

4. Given a GA with proportional selection and uniform crossover, but no mutation, applied to a linear fitness function, define
   $$D_t = f^* - \bar{f}(t)$$
   where f^* is the optimal fitness and $\bar{f}(t)$ is the average fitness of the population at time t. Show that the sequence $D_0, D_1, D_2, \ldots$ is a supermartingale. What can be concluded about this system?

5. Suppose we have a search space $\mathcal{X} = \{0, 1, 2\}$ and a fitness function $f(0) = 10$, $f(1) = 1$, $f(2) = 10$. Write a genetic algorithm for this problem using proportional selection and a 1% chance of an individual being mutated. If there is a mutation event then the individual is replaced by a random element of the search space. Run this algorithm, with a population size of 10, for a few hundred generations or more. What do you conjecture about the limiting distribution of this system? Now increase the population size to 200. What do you observe this time? Try even larger populations. (This exercise demonstrates the critical effect that population size can have on an evolutionary algorithm's behaviour.)

Chapter 6

The Dynamical Systems Model

In the last chapter we saw that the transition matrix for the genetic algorithm Markov chain is found by calculating the cumulative effects of selection, mutation and crossover. We now develop this construction in some detail, in order to examine the dynamics of a population as it moves from generation to generation. This will give us in a more complete form the exact model of Vose [293], which we introduced in Chapter 3.

6.1 Population Dynamics

Firstly, recall that we can represent a population as an incidence vector:

$$\boldsymbol{v} = (v_0, v_1, \ldots, v_{n-1})$$

in which v_k is the number of copies of individual $k \in \mathcal{X}$ in the population, so that

$$\sum_{k=0}^{n-1} v_k = N$$

where N is the population size. For simplicity, we identify the search space $\mathcal{X}$ with the set $\{0, 1, \ldots, n-1\}$. By using $\boldsymbol{v}$ as the means of representation, there is an implicit dependence on the population size. To obtain a more general representation, therefore, we consider a population to be described by the vector

$$\boldsymbol{p} = (p_0, p_1, \ldots, p_{n-1})$$

in which p_k is the *proportion* of individual $k \in \mathcal{X}$ that occurs in the population. Clearly

$$\sum_{k=0}^{n-1} p_k = 1.$$

Vectors of this kind can represent populations of any size. $\boldsymbol{p}$ is called a *population vector*. We can immediately recover the incidence vector from $\boldsymbol{p}$ for a given population size by $\boldsymbol{v} = N\boldsymbol{p}$. We can now view the action of a genetic algorithm as a trajectory of vectors $\boldsymbol{p} \in \mathbb{R}^n$. Population vectors are members of a subset of $\mathbb{R}^n$

$$\Lambda = \left\{ \boldsymbol{x} \in \mathbb{R}^n : x_k \geq 0 \text{ for all } k \text{ and } \sum_{k=0}^{n-1} x_k = 1 \right\},$$

which forms a *simplex*. When $n = 2$, the simplex is simply a straight line segment in the plane (see Figure 6.1), running from $(1,0)$ to $(0,1)$. The end-point $(1,0)$ represents a population comprising copies of individual 0 only. The other end-point $(0,1)$ represents populations with copies of individual 1 only. The points in between are populations containing a mixture of individuals. When $n = 3$, the simplex is a triangle embedded in $\mathbb{R}^3$ (see Figure 6.2). Again the vertices of the simplex represent uniform populations. When $n = 4$, the simplex is a tetrahedron embedded in $\mathbb{R}^4$, and so on up through higher dimensions. It should be noted that not all points in the simplex can represent populations. Points with irrational co-ordinates could never be populations. Populations of size N can only correspond to rational vectors with denominator N in each co-ordinate. However, as $N \to \infty$ the set of points corresponding to populations of size N becomes dense in the simplex, and the simplex is the closure of these points. Since the simplex is also bounded, this means that it is *compact* (see [293] for details). That is, all sequences of population vectors have a subsequence that converges to a point in the simplex (possibly irrational). When considering the dynamics of population vectors, it is therefore useful to include all points in the simplex.

In defining the transition matrix for the genetic algorithm, we saw that a new population is constructed by taking N samples from a distribution that depends on the previous generation, and on the operators that act upon it. In general terms, there will be a probability p'_k that individual k is sampled from this distribution. We can collect these probabilities into a vector

$$\boldsymbol{p}' = (p'_0, p'_1, \ldots p'_{n-1})$$

Note that $\boldsymbol{p}' \in \Lambda$. In other words, vectors in the simplex have a dual interpretation. They can represent the proportions of individuals in a pop-

6.1. POPULATION DYNAMICS

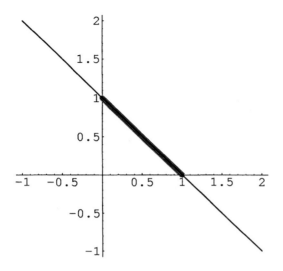

Figure 6.1: The simplex of possible populations when $n = 2$ forms a line-segment embedded in two-dimensional space.

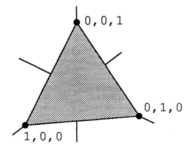

Figure 6.2: The simplex of possible populations when $n = 3$ forms a triangle embedded in a three-dimensional space.

ulation, and they can represent a probability distribution over individuals in the search space. We define the *generational operator*

$$\mathcal{G} : \Lambda \longrightarrow \Lambda$$

by $\mathcal{G}(p) = p'$ where p' is the probability distribution that is sampled to give the next population after p. As described previously, $\mathcal{G}(p)$ is sampled *multinomially* so that, if q is any population of size N, the probability that it is the next generation after p is

$$N! \prod \frac{(\mathcal{G}(p)_j)^{Nq_j}}{(Nq_j)!}$$

Thus $\mathcal{G}(p)$ describes the probability distribution over all populations for the next generation. It also has further interpretations, as in the following theorem.

Theorem 6.1 (Vose [295]) *If the current population is given by the population vector p, then the expected next population is $\mathcal{G}(p)$.*

One interpretation therefore, is that the vector $\mathcal{G}(p)$ describes the average over all possible next generations. Clearly, it is unlikely actually to *be* the next generation—it is only the *expected* value of the next population vector. There will be some variance about the average, and it can be shown that this variance is inversely proportional to the population size. That is, as N increases, the variance becomes smaller, and the probability distribution for the next generation is focused on $\mathcal{G}(p)$. Indeed, in the limit as $N \to \infty$ the variance shrinks to zero, and the next generation will actually be $\mathcal{G}(p)$. In the infinite population limit, the process becomes deterministic, and follows the trajectory in the simplex given by $p, \mathcal{G}(p), \mathcal{G}^2(p), \ldots$. The operator $\mathcal{G}$ can therefore also be interpreted as an *infinite population* model of a GA. For finite populations, of course, there will be a stochastic deviation from this trajectory, which can be viewed as arising from sampling errors.

A study of the dynamics of the genetic algorithm thus comprises an analysis of the operator $\mathcal{G}$, the trajectory it describes in Λ, and the effects of having a finite population creating deviations from this trajectory. This general theory has been worked out by Michael Vose and co-workers, and we shall follow their approach [293]. Some important issues will be:

1. Deriving equations for $\mathcal{G}$.

2. Relating the structural properties of the genetic operators to these equations.

6.2. SELECTION

3. Studying the dynamics of $\mathcal{G}$. We shall be interested in fixed points, and the trajectories of populations.

4. Relating the dynamics to the stochastic behaviour of finite populations.

We shall now look at the equations for $\mathcal{G}$ and their associated dynamics for selection, mutation, and crossover.

6.2 Selection

We have previously seen that if we use proportional selection, the probability of any individual $i \in \mathcal{X}$ being selected is

$$\mathbf{Pr}[i|v] = \frac{v_i f(i)}{\sum\limits_{j \in \mathcal{X}} v_j f(j)}$$

where v is the incidence vector for the population. Dividing the numerator and denominator of the fraction by the population size N enables us to re-write this in terms of the population vector p:

$$\mathbf{Pr}[i|p] = \frac{p_i f(i)}{\sum\limits_{j \in \mathcal{X}} p_j f(j)}$$

We can write this more compactly if we view the fitness function f as a vector $f \in \mathbb{R}^n$ given by $f_k = f(k)$. Viewing selection as an operator $\mathcal{F}: \Lambda \to \Lambda$ that acts on the simplex then

$$\mathcal{F}(p) = \frac{\text{diag}(f)p}{f^T p}$$

where $\text{diag}(f)$ is the diagonal matrix with entries from vector f along the diagonal, and $f^T p$ is the inner product of vectors f and p.

For example, suppose our search space is $\mathcal{X} = \{0, 1, 2, 3\}$ and the fitness function is

$$\begin{aligned} f(0) &= 2 \\ f(1) &= 1 \\ f(2) &= 3 \\ f(3) &= 2 \end{aligned}$$

We identify the function f with the vector $\boldsymbol{f} = (2,1,3,2)$. The operator $\mathcal{F}$ is then given by:

$$\mathcal{F}(\boldsymbol{p}) = \frac{1}{\boldsymbol{f}^T\boldsymbol{p}} \begin{bmatrix} 2 & 0 & 0 & 0 \\ 0 & 1 & 0 & 0 \\ 0 & 0 & 3 & 0 \\ 0 & 0 & 0 & 2 \end{bmatrix} \boldsymbol{p}$$

where

$$\boldsymbol{f}^T\boldsymbol{p} = \sum_k f_k p_k = \sum_k f(k) p_k$$

is the average fitness of the population represented by $\boldsymbol{p}$.

If proportional selection is the only genetic operator, then $\mathcal{G} = \mathcal{F}$ and we now have the evolutionary equation for $\mathcal{G}$. We can therefore find the dynamics under selection. Firstly, we establish the fixed points. A fixed point is a point $\boldsymbol{v}$ such that $\mathcal{G}(\boldsymbol{v}) = \boldsymbol{v}$. Let $\boldsymbol{e}_k$ be the vector containing a 1 in position k and zeros elsewhere. Such a vector is one the vertices of the simplex and represents a *uniform population*. It is clear from the definition of $\mathcal{F}$ that each such vertex is a fixed point of proportional selection. (Recall that uniform populations are also absorbing states in the Markov chain.) However, these are not the only fixed points: it is simple to prove the following:

Theorem 6.2 *Let $A \subseteq \mathcal{X}$ be a set of individuals in the search space all having the same fitness value. The set*

$$\mathcal{H}\{\boldsymbol{e}_k : k \in A\} = \left\{ \sum_{k \in A} \alpha_k \boldsymbol{e}_k : \alpha_k \geq 0 \text{ for all } k \text{ and } \sum \alpha_k = 1 \right\}$$

is called the convex hull *of the vertices. All points in this set are fixed points of proportional selection.*

The fixed points of a GA that uses only proportional selection are thus the uniform populations, and any mixed population containing individuals with the same fitness value. To continue our example, consider the set $A = \{0, 3\} \subseteq \mathcal{X}$. With the given fitness function, $f(0) = f(3)$. The convex hull of $\{\boldsymbol{e}_0, \boldsymbol{e}_3\}$ is

$$\mathcal{H}\{\boldsymbol{e}_0, \boldsymbol{e}_3\} = \{\alpha_0 \boldsymbol{e}_0 + \alpha_3 \boldsymbol{e}_3 : \alpha_0, \alpha_3 \geq 0 \text{ and } \alpha_0 + \alpha_3 = 1\}$$

and any vector in this set is a fixed point of $\mathcal{F}$, since

$$\mathcal{F}(\alpha_0 \boldsymbol{e}_0 + \alpha_3 \boldsymbol{e}_3) = \frac{1}{2\alpha_0 + 2\alpha_3} \begin{bmatrix} 2 & 0 & 0 & 0 \\ 0 & 1 & 0 & 0 \\ 0 & 0 & 3 & 0 \\ 0 & 0 & 0 & 2 \end{bmatrix} (\alpha_0 \boldsymbol{e}_0 + \alpha_3 \boldsymbol{e}_3)$$

6.2. SELECTION

$$= \frac{2\alpha_0 e_0 + 2\alpha_3 e_3}{2(\alpha_0 + \alpha_3)}$$
$$= \alpha_0 e_0 + \alpha_3 e_3$$

However, for *finite* populations, only the uniform populations are absorbing states of the Markov process. This is because mixed populations will fluctuate randomly until eventually a uniform population is reached. This effect is sometimes called *genetic drift*.

We can now find the trajectory of $\mathcal{G}$ in the simplex by proving the following simple theorem. A form of this theorem appears in [284].

Theorem 6.3 *Suppose $\mathcal{G}$ is defined by*

$$\mathcal{G}(p) = \frac{diag(f)p}{f^T p}$$

Let $p(t)$ be the population at time t, i.e., $p(t) = \mathcal{G}^t(p(0))$. Then

$$p(t)_i = \frac{f_i^t p(0)_i}{\sum_j f_j^t p(0)_j}$$

Proof By induction. The case $t = 0$ is trivial:

$$\frac{f_i^0 p(0)_i}{\sum_j f_j^0 p(0)_j} = \frac{p(0)_i}{\sum_j p(0)_j} = p(0)_i$$

Now assume that the hypothesis holds for $t - 1$. Then

$$p(t)_i = \frac{f_i p(t-1)_i}{\sum_j f_j p(t-1)_j}$$

by definition of $\mathcal{G}$. By induction hypothesis, the numerator is

$$f_i \frac{f_i^{t-1} p(0)_i}{\sum_j f_j^{t-1} p(0)_j} = \frac{f_i^t p(0)_i}{\sum_j f_j^{t-1} p(0)_j}$$

while the denominator is

$$\sum_j f_j \frac{f_j^{t-1} p(0)_j}{\sum_k f_k^{t-1} p(0)_k} = \sum_j \frac{f_j^t p(0)_j}{\sum_k f_k^{t-1} p(0)_k}$$

and therefore

$$p(t)_i = \frac{f_i^t p(0)_i}{\sum_j f_j^t p(0)_j}$$

as required. □

To illustrate these ideas, let us take a simpler example, with just three individuals in the search space. This will allow us to draw what is happening in the simplex. Suppose the fitness function is

$$f(0) = 3$$
$$f(1) = 2$$
$$f(2) = 1$$

We can calculate the flow of $\mathcal{G}$ in the simplex under proportional selection. This is illustrated in Figure 6.3, which also shows a particular trajectory for a given starting point $\boldsymbol{p}(0) = (0.01, 0.24, 0.75)$. This initial population is fairly close to the vertex corresponding to the worst solution. As each generation goes by, the population moves firstly towards the second best vertex and then turns towards the optimal vertex, where it converges. We can use our previous theorem to calculate exactly the proportion of each solution in the population at each time-step of this trajectory. The results are displayed in Figure 6.4. Notice how the curve for the second vertex rises and falls as the population approaches it and then turns towards the optimum.

6.2.1 Ranking selection

Fitness-proportional selection is just one of several selection techniques used in genetic algorithms. Others, such as rank and tournament selection, can also be described as operators that act upon population vectors [293]. We shall describe the operator $\mathcal{F}$ for a particular form of rank selection.

With a population of size N, there are N possible ranks that can be assigned (ties being broken in some way). For each of these possible ranks, we assign a probability that the associated population member is selected. Let the probability that an individual with rank i is selected be $R(i)$, for $i = 0, 1, \ldots N-1$, where rank 0 is the worst and rank $N-1$ the best. Now we define a probability density function $\rho(x)$ to be the following step function:

$$\rho(x) = N.R(\lfloor Nx \rfloor), \quad 0 \leq x \leq 1,$$

where $\lfloor \cdot \rfloor$ is the *floor* function. Note that

$$\int_0^1 \rho(x) dx = 1$$

6.2. SELECTION

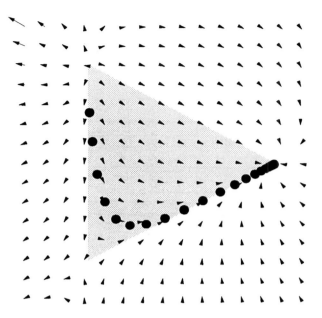

Figure 6.3: The flow of $\mathcal{G}$ on the simplex (light grey triangle) under proportional selection. A particular trajectory is plotted, which converges to the fixed point at the optimal vertex.

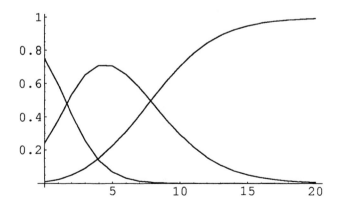

Figure 6.4: The proportion of each solution in the population for the trajectory shown in Figure 6.3. The worst individual is quickly eradicated. The second-best individual grows in popularity as the population heads towards the corresponding vertex. The population then swings towards the optimal vertex. The second-best individual dies away and the optimal solution fills the whole population.

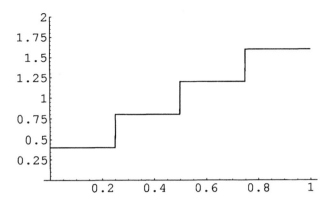

Figure 6.5: Probability density function $\rho(x)$ corresponding to a particular set of ranking probabilities $R = (0.1, 0.2, 0.3, 0.4)$ with population size $N = 4$. The density function is defined to be $\rho(x) = 4R(\lfloor 4x \rfloor)$.

For example, suppose that $N = 4$, and we assign the following probabilities to ranks

i	0	1	2	3
$R(i)$	0.1	0.2	0.3	0.4

then the step function $\rho(x)$ is as shown in Figure 6.5.

Given a population vector p, suppose that $k \in \mathcal{X}$ is the worst individual in p, i.e., p_k is non-zero and $p_j = 0$ for all $j \in \mathcal{X}$ such that $f(j) < f(k)$. The probability that k is selected is given by collecting together the first Np_k steps of the function ρ, this being the number of copies of k that are in the population. The probability of selecting the next best individual is found in a similar way, by collecting together the appropriate number of steps of ρ. Notice that this works even if there are no copies of an individual in the population, as then we simply collect zero steps. In general, then, we can find the probability of selecting any individual $k \in \mathcal{X}$ by integrating over ρ to collect together the appropriate number of steps:

$$\mathcal{F}(p)_k = \int_{\sum_{[f(j)<f(k)]} p_j}^{\sum_{[f(j)\leq f(k)]} p_j} \rho(y) dy$$

Continuing with the example, suppose we have a search space with four individuals ($n = 4$) and a fitness function such that $f(0) < f(1) < f(2) < f(3)$, and the current population vector is $p = (1/4, 1/2, 0, 1/4)$. The following table shows how the rank probabilities are calculated:

k	$\sum[f(j) < f(k)]p_j$	$\sum[f(j) \leq f(k)]p_j$	$\mathcal{F}(p)_k$
0	0	1/4	$\int_0^{1/4} \rho(y)dy = 0.1$
1	1/4	1/4 + 1/2	$\int_{1/4}^{3/4} \rho(y)dy = 0.2 + 0.3 = 0.5$
2	1/4 + 1/2	1/4 + 1/2 + 0	$\int_{3/4}^{3/4} \rho(y)dy = 0$
3	1/4 + 1/2 + 0	1/4 + 1/2 + 0 + 1/4	$\int_{3/4}^{1} \rho(y)dy = 0.4$

This method of defining rank-based selection can be generalised by using any non-decreasing probability density function $\rho(x)$ in the integral.

6.3 Mutation

In the previous chapter, the contribution of mutation to the transition matrix was described in terms of a $n \times n$ matrix U in which $U_{i,j}$ is the probability that individual $j \in \mathcal{X}$ mutates to individual $i \in \mathcal{X}$. The same matrix directly gives the effect of mutation on a population vector. We seek an operator $\mathcal{U} : \Lambda \to \Lambda$ such that $\mathcal{U}(p)$ is a vector describing the probability distribution over $\mathcal{X}$, from which the next population is chosen multinomially—reflecting the effect of applying mutation to the population vector p. It is simple to show that

$$\mathcal{U}(p) = Up$$

is the required operator.

The combined effects of applying selection and mutation to a population vector are given by $\mathcal{U} \circ \mathcal{F}$. For the case of proportional selection this gives

$$\mathcal{U} \circ \mathcal{F}(p) = \frac{U \operatorname{diag}(f) p}{f^T p}.$$

If our GA consists only of proportional selection and mutation (i.e. no crossover) then $\mathcal{G} = \mathcal{U} \circ \mathcal{F}$. The fixed points of $\mathcal{G}$ in this case are given by

$$\frac{U \operatorname{diag}(f) p}{f^T p} = p,$$

so that

$$U \operatorname{diag}(f) p = (f^T p) p.$$

For a matrix A, any scalar value λ such that $Ax = \lambda x$ for some x is called an *eigenvalue* of the matrix A. The vector x is called an *eigenvector* corresponding to λ. (Note that if $\boldsymbol{x}$ is an eigenvector corresponding to a particular eigenvalue λ, so also is the vector $\alpha\boldsymbol{x}$ for any scalar α.) We conclude that fixed points of $\mathcal{G}$, defined as proportional selection and mutation, are eigenvectors of the matrix $\boldsymbol{U}\mathrm{diag}(\boldsymbol{f})$, scaled so that the entries of the eigenvector sum to one. The eigenvalue to which an eigenvector $\boldsymbol{p}$ corresponds is $\boldsymbol{f}^T\boldsymbol{p}$—the average fitness of fixed point population $\boldsymbol{p}$.

It is easy to show that the converse is also true: any suitably scaled eigenvector of $\boldsymbol{U}\mathrm{diag}(\boldsymbol{f})$ is a fixed point of $\mathcal{G}$.. This means that we can find all the fixed points by calculating the eigensystem of $\boldsymbol{U}\mathrm{diag}(\boldsymbol{f})$ and scaling the eigenvectors so that their components sum to one. The eigenvalues will give us the average fitness of these fixed points. It should be noted, however, that such vectors will not necessarily correspond to actual finite populations. In the first place they might be irrational. Even if they are rational, they may not have a common denominator N. Moreover, they may contain negative components or even complex ones. The eigenvectors represent fixed points of the infinite population limit, both inside and outside the simplex. However, we can guarantee that exactly one of these fixed points will lie inside the simplex, thanks to a more general version of the Perron-Frobenius theorem, which we met in the previous chapter. This theorem tells us that, as long as $\boldsymbol{U}\mathrm{diag}(\boldsymbol{f})$ has only positive entries, it has exactly one eigenvector in Λ. This eigenvector corresponds to the leading eigenvalue (the one with the largest absolute value) and this eigenvalue is guaranteed to be positive. This fixed point, then, is the unique point to which all trajectories in Λ converge: it is *asymptotically stable*.

Even though the dynamics of $\mathcal{G}$ provides the infinite population limit of a genetic algorithm, the fixed points (inside and outside Λ) are important in understanding the behaviour of a finite population. Suppose we define the *force* of $\mathcal{G}$ at a point $\boldsymbol{p}$ to be

$$\|\mathcal{G}(\boldsymbol{p}) - \boldsymbol{p}\|,$$

i.e., it is the distance that $\boldsymbol{p}$ moves under $\mathcal{G}$. Obviously, at a fixed point the force is zero. Now suppose that $\boldsymbol{p}$ is a population vector corresponding to some population of size N, and that it is near to a fixed point of $\mathcal{G}$. Because $\mathcal{G}$ is continuous, the force of $\mathcal{G}$ at $\boldsymbol{p}$ will be very small. This means that the expected next population will be very close to $\boldsymbol{p}$. But for small population sizes, there may be very few population vectors near to $\boldsymbol{p}$. The genetic algorithm will, with high probability, stay at or close to $\boldsymbol{p}$. In this case $\boldsymbol{p}$ is not, properly speaking, a fixed point of the genetic algorithm, and we call it a *metastable* state. Metastable populations have a relatively high

6.3. MUTATION

probability of occurring in the limiting distribution of the corresponding Markov process. In fact, fixed points just outside Λ can create metastable states, since they will tend to trap populations in Λ that come close to them. We often find in practice that a finite population will spend long periods of time in such states. Exactly which metastable state is first encountered depends on the initial population, and on the stochastic nature of the system. The larger the population size, the less the stochasticity, and the more the behaviour will approximate the infinite population limit.

6.3.1 An example

We can calculate the fixed points of a genetic algorithm with proportional selection and bit-wise mutation on the following fitness function:

$$\begin{align} f(0\,0\,0) &= 10 \\ f(0\,0\,1) &= 1 \\ f(0\,1\,0) &= 1 \\ f(0\,1\,1) &= 1 \\ f(1\,0\,0) &= 1 \\ f(1\,0\,1) &= 1 \\ f(1\,1\,0) &= 1 \\ f(1\,1\,1) &= 11 \end{align}$$

Clearly, this function has two peaks, the optimal one (at 1 1 1) being only slightly larger than the other (at 0 0 0). This function could be described as 'two needles in a haystack'. Taking the mutation rate to be $\mu = 0.05$, we calculate the mutation matrix

$$U_{i,j} = \mu^d (1-\mu)^{(3-d)}$$

where $d = d_H(i,j)$ is the Hamming distance between the strings i and j. Calculating the eigenvectors of $U \operatorname{diag}(f)$ confirms that there is exactly one inside the simplex, namely

$(0.00224, 0.0032, 0.0032, 0.04876, 0.0032, 0.04876, 0.04876, 0.84188)$,

which we see is concentrated around the global maximum. The corresponding eigenvalue is 9.439, which is the average fitness of this population. However, there is another eigenvector which is very close to the simplex:

$(0.84479, 0.04941, 0.04941, 0.00304, 0.04941, 0.00304, 0.00304, -0.00214)$

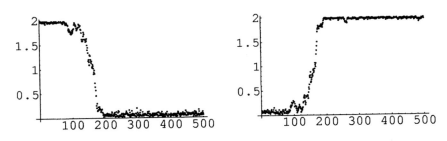

Figure 6.6: Distance of the population in a running GA from the stable fixed point within the simplex (left) and the metastable point just outside the simplex (right).

This is close to the suboptimal point. The argument above tells us that any finite population that is close to either of these fixed points is likely to stay there for a considerable time. In the infinite population limit, of course, all populations will converge on the first of these. The existence of the second one so close to the simplex however has an important influence on finite population behaviour. Running a GA on this problem with a population size of 200 illustrates this effect. We start with an initial population close to the second fixed point by taking 160 copies of 0 0 0 and 20 each of 0 0 1 and 0 1 0, giving a population vector $(0.8, 0.1, 0.1, 0, 0, 0, 0, 0)$. On a single run, we plot the distance between the current population and each of the two fixed points. We measure the distance between two vectors by

$$d(\boldsymbol{x}, \boldsymbol{y}) = \sum_j |x_j - y_j|$$

The results of a typical run are show in Figure 6.6. We can see that the population stays close to the metastable region for approximately 200 generations. This result will vary from run to run owing to the stochastic nature of the system.

6.3.2 Diagonalization

The maximum number of eigenvalues that a $n \times n$ matrix $\boldsymbol{A}$ can have is n. If $\boldsymbol{A}$ has a full set of n distinct eigenvalues, then the corresponding eigenvectors are linearly independent. If we create a matrix $\boldsymbol{Z}$ whose columns are the eigenvectors of $\boldsymbol{A}$, then it can be shown that $\boldsymbol{Z}^{-1}\boldsymbol{A}\boldsymbol{Z}$ is a diagonal matrix, whose diagonal entries are the corresponding eigenvalues. $\boldsymbol{A}$ is said to be *diagonalizable*. The set of eigenvectors forms a basis for the vector space on which $\boldsymbol{A}$ operates and, in this basis, $\boldsymbol{A}$ becomes a diagonal matrix.

6.3. MUTATION

Now suppose that $U\text{diag}(f)$ is diagonalizable, i.e., we have found all the eigenvalues and they are distinct. It would seem that $\mathcal{G}$ should be naturally described in the basis of eigenvectors. Let Z be the matrix containing the eigenvectors as columns, so that $Z^{-1}U\text{diag}(f)Z = \text{diag}(\lambda)$ where λ is a vector containing all the eigenvalues. Given any vector x, we can express it in the new eigenvector basis as $Z^{-1}x$. So transforming the evolutionary equation into this basis we obtain the following:

$$\begin{aligned} Z^{-1}\mathcal{G}(p) &= Z^{-1}\frac{U\text{diag}(f)p}{f^T p} \\ &= Z^{-1}\frac{Z\text{diag}(\lambda)Z^{-1}p}{f^T p} \\ &= \frac{\text{diag}(\lambda)Z^{-1}p}{f^T p} \end{aligned}$$

To proceed further, we use an alternative expression for $f^T p$. Let $h : \mathbb{R}^n \to \mathbb{R}$ be a function of vectors which adds up their component values:

$$h(x) = \sum_j x_j,$$

so that $p \in \Lambda$ implies that $h(p) = 1$. Then

$$f^T p = h(\text{diag}(f)p)$$

The following lemma will be useful:

Lemma 6.4 *Let A be a matrix whose columns sum to 1. That is $\sum_i A_{i,j} = 1$ for each j. Then $h(Ax) = h(x)$ for all x.*

Proof

$$h(Ax) = \sum_i (Ax)_i = \sum_i \sum_j A_{i,j} x_j = \sum_j x_j \sum_i A_{i,j} = \sum_j x_j = h(x)$$

$\square$

Now the mutation matrix U has the property that its columns sum to one, as does the matrix Z if we have scaled the eigenvectors appropriately. Therefore

$$\begin{aligned} f^T p &= h(\text{diag}(f)p) \\ &= h(U\text{diag}(f)p) \\ &= h(Z\text{diag}(\lambda)Z^{-1}p) \\ &= h(\text{diag}(\lambda)Z^{-1}p) \\ &= \lambda^T Z^{-1} p \end{aligned}$$

Putting everything together, suppose we have a population vector $p(t)$ at time t, so that
$$p(t+1) = \mathcal{G}(p(t))$$
Let $q(t) = Z^{-1}p(t)$ be the same population expressed in the new basis. Then we have

$$\begin{aligned} q(t+1) &= Z^{-1}\mathcal{G}(p(t)) \\ &= \frac{\text{diag}(\lambda)Z^{-1}p(t)}{f^T p(t)} \\ &= \frac{\text{diag}(\lambda)Z^{-1}p(t)}{\lambda^T Z^{-1}p} \\ &= \frac{\text{diag}(\lambda)q(t)}{\lambda^T q(t)} \end{aligned}$$

Clearly, this equation is identical to the one that we derived for proportional selection only! The only difference is that now we have a 'fitness' vector λ which contains the eigenvalues of $U\text{diag}(f)$. This means that the dynamics of the infinite population genetic algorithm under mutation and fitness-proportional selection with fitness vector f is identical, in the new basis, to the dynamics of proportional selection on its own with fitness vector λ. But we already know how to solve the dynamics of proportional selection. We can apply the same result now:

$$q(t)_i = \frac{\lambda_i^t q(0)_i}{\sum_j \lambda_j^t q(0)_j}$$

remembering that $q = Z^{-1}p$ represents the population in the eigenvector basis.[1].

We can use this result to study the infinite population dynamics for the system in the previous example (the 'two needles in a haystack' problem). Transforming co-ordinates to the Z basis, we plot the proportion of each of the two eigenvectors (optimal and metastable) found at each generation, starting with the same initial population as before. The results are shown in Figure 6.7. It can be seen that the population is expected to switch from one to the other between generations 50 and 100. Remember, though, that this is for the infinite population. The effect of having a finite population size is to make the population stall in the region of the metastable state for a much longer time, as the experimental results in Figure 6.6 show.

---

[1]A form of this result appears in [284]

6.3. MUTATION

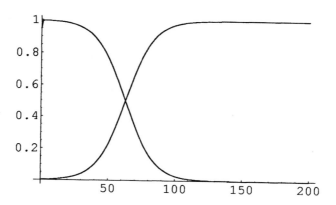

Figure 6.7: The proportion of the optimal and metastable eigenvectors in a population started at the same initial point as in Figure 6.6. In the infinite population limit it can be seen that the population should switch from one to the other between generations 50 and 100. Actual finite populations will typically take longer than this to switch.

What exactly does this change of basis mean? We can illustrate what is happening by returning to an earlier example:

$$f(0) = 3$$
$$f(1) = 2$$
$$f(2) = 1$$

We have already seen that with no mutation, the fixed points are the vertices of the simplex. We now add mutation. Suppose there is a probability μ that an individual will mutate. If it does so, then it becomes one of the other individuals with equal probability, generating the mutation matrix

$$U = \begin{bmatrix} 1-\mu & \mu/2 & \mu/2 \\ \mu/2 & 1-\mu & \mu/2 \\ \mu/2 & \mu/2 & 1-\mu \end{bmatrix}.$$

The effect on the dynamics of setting $\mu = 0.05$ and $\mu = 0.1$ are shown in Figure 6.8. We observe that as mutation increases, so one of the fixed points moves inside the simplex (this is the leading eigenvector, as predicted by Perron-Frobenius) while the others move out. The extent to which they move depends on the mutation rate: the higher the rate, the further they move. Fixed points staying close to the simplex may form metastable states within it. Note that the three fixed points form a triangle. The flow within this triangle looks rather like the flow under selection in the simplex, but for a different fitness function—the function comprising the eigenvalues of $U\text{diag}(f)$. This is what the transformation Z means. It maps each of the fixed points back to the vertices of the simplex, and simultaneously maps the flow of the dynamical system inside the fixed point triangle into a flow in the simplex. This flow corresponds to a selection-only flow with the new fitness function.

6.4 Crossover

We now seek to define the action of crossover on a population vector. That is, we wish to define an operator $\mathcal{C} : \Lambda \to \Lambda$ such that the kth component of $\mathcal{C}(p)$ is the probability that individual $k \in \mathcal{X}$ results from applying crossover to population p. Suppose $i, j \in \mathcal{X}$ and that the probability that these two cross to form k is $r(i, j, k)$, as sketched previously in Chapter 3. For example, with uniform crossover defined on bitstrings of length 3,

$$r(0\,0\,0, 0\,1\,1, 0\,0\,1) = 0.25$$

The probability that k is created by applying crossover to a population p is then found by summing over all the possible ways this can happen:

$$\mathcal{C}(p)_k = \sum_{i,j} p_i p_j r(i, j, k)$$

Sometimes crossover is symmetric, so that $r(i, j, k) = r(j, i, k)$, as with uniform crossover, for example. However, even if crossover is not symmetric, we can define symmetric matrices M_k having i, jth component

$$\frac{1}{2}\Big(r(i, j, k) + r(j, i, k)\Big)$$

It is straightforward then to check that

$$\mathcal{C}(p)_k = p^T M_k p$$

Each component of $\mathcal{C}$ is a quadratic form, and $\mathcal{C}$ itself is a quadratic operator.

As an example, consider bitstrings of length 2, and one-point crossover, where we allow crosspoints before the first and after the last bit[2]. We

---

[2] We recognize that this choice of crosspoints is rare in practice, as it leads to cloning, but we need it for this illustration. Otherwise we would need longer strings and the matrices would be rather large!

6.4. CROSSOVER

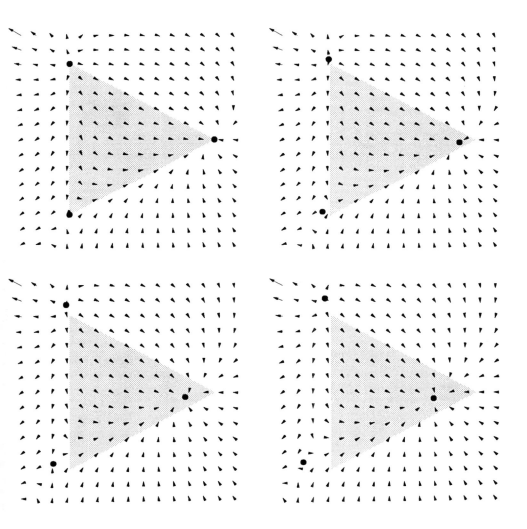

Figure 6.8: The effect of mutation on the flow of $\mathcal{G}$. Mutation rate is $\mu = 0$ (top-left), $\mu = 0.05$ (top-right), $\mu = 0.1$ (bottom-left), $\mu = 0.15$ (bottom-right).

associate the search space $\{0\,0, 0\,1, 1\,0, 1\,1\}$ with the integers $\{0, 1, 2, 3\}$ using the standard binary representation. $r(i, j, \mathbf{0})$ therefore is the probability that i and j cross to form $\mathbf{0} = (0\,0)$. The following table gives the different values of $r(i, j, \mathbf{0})$:

	0 0	0 1	1 0	1 1
0 0	1	1/3	2/3	1/3
0 1	2/3	0	1/3	0
1 0	1/3	0	0	0
1 1	1/3	0	0	0

It can be seen that one-point crossover is not symmetric since, for example,

$$r(0\,0, 0\,1, 0\,0) \neq r(0\,1, 0\,0, 0\,0)$$

However, we can take the matrix $\boldsymbol{M}_0$ to be symmetric and still obtain the correct quadratic form for $\mathcal{C}$.

$$\boldsymbol{M}_0 = \begin{bmatrix} 1 & 1/2 & 1/2 & 1/3 \\ 1/2 & 0 & 1/6 & 0 \\ 1/2 & 1/6 & 0 & 0 \\ 1/3 & 0 & 0 & 0 \end{bmatrix}$$

Doing this for each $k \in \mathcal{X}$ gives us

$$\mathcal{C}(\boldsymbol{p}) = (\boldsymbol{p}^T \boldsymbol{M}_0 \boldsymbol{p}, \boldsymbol{p}^T \boldsymbol{M}_1 \boldsymbol{p}, \ldots, \boldsymbol{p}^T \boldsymbol{M}_{n-1} \boldsymbol{p})$$

where each $\boldsymbol{M}_k$ is symmetric.

It should be noticed that this means that different crossovers might lead to the same operator $\mathcal{C}$, indicating that they have identical effects on a population. For example, suppose we have two different forms of one-point crossover. The first one generates a single offspring from the left-hand end of one parent and the right-hand end of the other parent. The second one generates both offspring and selects one of them at random. It can be checked that both of these crossovers have the same corresponding set of symmetric matrices $\boldsymbol{M}_k$ and that therefore their effect on populations is identical. (A similar effect in the case of permutations is demonstrated by Whitley and Yoo [309].)

6.4.1 Mixing

If we combine crossover with mutation we obtain the *mixing* operator

$$\mathcal{M} = \mathcal{C} \circ \mathcal{U}$$

6.4. CROSSOVER

where mutation is applied to parents before crossover. For a large class of *independent* mutation operators (including all the standard ones on bit-strings), it doesn't matter whether mutation is applied before crossover or afterwards [293]. The mixing operator also turns out to be a quadratic operator [243]:

$$\mathcal{M}(p)_k = p^T(U^T M_k U)p,$$

where U is the mutation matrix. Some of the properties of $\mathcal{M}$ that relate to the structure of the search space $\mathcal{X}$ will be described in the following section.

The full sequence of selection, mutation and crossover gives us the complete operator for the genetic algorithm:

$$\mathcal{G} = \mathcal{M} \circ \mathcal{F}$$

If selection is proportional to fitness then, as we have seen,

$$\mathcal{F}(p) = \frac{\text{diag}(f)p}{f^T p}.$$

One of the properties of quadratic operators such as $\mathcal{M}$ is that, for any vector x and scalar α

$$\mathcal{M}(\alpha x) = \alpha^2 \mathcal{M}(x).$$

From this it follows that, for proportional selection

$$\mathcal{M}\left(\frac{\text{diag}(f)p}{f^T p}\right) = \frac{\mathcal{M}(\text{diag}(f)p)}{(f^T p)^2}$$

and so we obtain

$$\mathcal{G}(p)_k = \frac{p^T(\text{diag}(f)U^T M_k U \text{diag}(f))p}{(f^T p)^2}$$

for each $k \in \mathcal{X}$, which gives us the complete equation for the genetic algorithm. Iterating this equation gives us the infinite population trajectory for the genetic algorithm with proportional selection.

As in the case of selection and mutation, the dynamics of a finite population genetic algorithm with crossover are strongly influenced by the fixed points of the infinite population system [290]. It can be shown empirically that in many cases the finite population spends most of its time in the vicinity of a fixed point. It eventually escapes and quickly becomes captive to another. It sometimes happens that whole regions of the simplex can stall the population. If the force of $\mathcal{G}$ is small on a connected subset then the population will drift randomly around it until it manages to escape. Such regions are often referred to as *neutral networks*, as the average fitness of the population does not alter significantly within them, and it appears that they may constitute manifolds of the infinite population fixed points. However, there is as yet no known method for analytically determining the fixed points of a genetic algorithm with crossover, and so most investigations of this phenomenon are either empirical, or relate to carefully constructed fitness functions. It should be noted that the fixed points may well contain complex numbers. These may still be 'close' to the simplex, if our notion of distance is extended appropriately. They are just as important as real-valued fixed points in understanding the dynamics of the system.

Analysing the properties of $\mathcal{G}$ is difficult and there are many open questions. Two important issues concern the convergence of iterates of $\mathcal{G}$. We shall say that $\mathcal{G}$ is *focused* if it is continuously differentiable and its iterates always converge to a fixed point. The differentiability of $\mathcal{G}$ is not a problem, but little is known about conditions under which the iterates converge (as opposed to forming a periodic orbit, for example). It is known that, when there is no crossover, convergence to a fixed point is assured. With crossover, the situation becomes more complicated. When the search space is $\{0,1\}^{\ell}$, and in the case of linear fitness functions and no mutation, it is known that $\mathcal{G}$ is focused [289]. A perturbation argument shows that this continues to be the case for small mutation rates and nearly linear fitness functions. However, it is possible to define pathological forms of mutation (which are not used in practice) for which periodic attractors are found [315]. Certain forms of 'self-adaptive' mutation schemes can produce periodic or even chaotic behaviour [317]. Finding precise conditions for $\mathcal{G}$ to be focused is thus an important open question.

The second issue concerns whether or not the fixed points are *hyperbolic*. A fixed point x is said to be hyperbolic if the differential of $\mathcal{G}$ at the point x has no eigenvalue with modulus equal to 1. Hyperbolicity is an important concept in determining the stability of fixed points. It is known that for the search space of binary strings, with proportional selection, bitwise mutation (at a positive rate) and any crossover, fixed points are hyperbolic for almost all fitness functions.[3] However, it is an important open issue to generalise this result to other search spaces, and to the operators that act upon them.

---

[3]More precisely, the set of functions for which this statement is true is dense in the set of all fitness functions.

6.5 Representational Symmetry

In this section we shall investigate the relationships that exist between structures within a search space and the operators (crossover and mutation) that act upon them. These properties of $\mathcal{M}$ will be analysed from a purely syntactic viewpoint, by considering the way in which crossover and mutation act on structural properties of the representations of points in the search space. Whether or not these operations make sense semantically (that is, in terms of the fitness of the points) is a difficult and fundamental open question. At present, the best we can say is that if our knowledge of the search problem at hand leads us to think that a certain representation is sensible, then crossover and mutation can act on that representation in certain interesting ways. Moreover, this can lead to a simplification of the equations describing the dynamics of the infinite population model.

Suppose we have selected a particular representation in order to obtain our search space $\mathcal{X}$. There may be various structures and symmetries implicit in this representation. Indeed, if there is no such structure, searching the space will amount to random sampling. An example of such a structure is the notion of a *landscape*, which we shall consider in greater detail in Chapter 9. We might choose a representation so that points of the search space correspond to vertices in a graph. Neighbouring vertices (joined by an edge) correspond to points in the search space that are similar in some sense. For example, the set of fixed-length binary strings can be thought of as forming a graph in which neighbours are points that are Hamming distance 1 apart. We can model such structures and their symmetries by considering *permutations* of the search space (i.e., bijections acting on $\mathcal{X}$) that preserve the structural properties in which we are interested. In the case of the landscape graph, such a permutation would have to map vertices in such a way that if two vertices were connected by an edge, then so would be the images of these vertices. The collection of such permutations represent the different ways in which points of the search space could be relabelled, whilst preserving the essential structural properties.

These structural symmetries may be describable mathematically in terms of a *group* that acts on the search space [244]. We denote this group by $L(\mathcal{X})$. It comprises a set of permutations of $\mathcal{X}$ which satisfies the following conditions:

1. The identity permutation is in L.

2. If $a \in L$ then the inverse permutation a^{-1} is also in L.

3. If a and b are in L, then so is $a \circ b$ (where $\circ$ is the group operation).

We shall assume throughout that this group acts *transitively* on $\mathcal{X}$, that is, for any $i, j \in \mathcal{X}$ there exists some $a \in L(\mathcal{X})$ such that $a(i) = j$.

A special case of a group acting on $\mathcal{X}$ is when $\mathcal{X}$ itself forms a group. For example, the set of fixed-length binary strings forms a group under the operation of bitwise addition modulo 2. A second example is when $\mathcal{X}$ is a set of permutations (as in the travelling salesman problem), for which the group operator is composition.

We can identify permutations of $\mathcal{X}$ with permutation matrices that act on Λ. For each $a \in L(\mathcal{X})$ we may define a matrix $\sigma_a : \Lambda \to \Lambda$ by

$$\sigma_a(x_0, \ldots, x_{n-1}) = (x_{a^{-1}(0)}, \ldots, x_{a^{-1}(n-1)}).$$

It is easy to check that this is equivalent to defining

$$(\sigma_a)_{i,j} = [i = a(j)].$$

Note also that the set of matrices $\{\sigma_a : a \in L(\mathcal{X})\}$ forms a group under matrix multiplication that is isomorphic to $L(\mathcal{X})$.

We say that the mixing operator $\mathcal{M}$ *commutes* with $L(\mathcal{X})$ if $\mathcal{M} \circ \sigma_a = \sigma_a \circ \mathcal{M} \;\forall a \in L(\mathcal{X})$. When this happens, the definition of $\mathcal{M}$ simplifies considerably. Recall that $\mathcal{M}$ is defined by n matrices:

$$\mathcal{M}(x)_k = x^T M_k x.$$

If mixing commutes with $L(\mathcal{X})$ then, for each $k \in \mathcal{X}$, let $a \in L(\mathcal{X})$ be a permutation such that $a(0) = k$. Then

$$M_k = M_{a(0)} = \sigma_a M_0 \sigma_a^T$$

In other words, the probabilities in each matrix M_k are exactly those that are in M_0, but shuffled around according to the permutation matrix σ_a. We have already touched on the benefits of this property in Chapter 3: the relationship between mixing (crossover and mutation) and the structural symmetries of $\mathcal{X}$ lead to a simplification in the mathematical expression for $\mathcal{M}$. All the relevant probabilities are stored in one matrix M_0 which is called the *mixing matrix*. We therefore only have to calculate this matrix in order to capture the whole effect of crossover (and mutation).

What kinds of crossover and mutation have this property? From the definition of crossover, it is clear that the relation

$$r(a(i), a(j), a(k)) = r(i, j, k)$$

6.5. REPRESENTATIONAL SYMMETRY

will give this result. That is, the action of crossover must be invariant with respect to the relabellings which preserve the structures of the search space. Similarly, for mutation we require

$$U_{a(i),a(j)} = U_{i,j}$$

We can now continue with the example of $\mathcal{X} = \{00, 01, 10, 11\}$ in the case of 1X. We have already calculated M_0. The complete set of matrices for this crossover operator is

$$M_0 = \begin{bmatrix} 1 & 1/2 & 1/2 & 1/3 \\ 1/2 & 0 & 1/6 & 0 \\ 1/2 & 1/6 & 0 & 0 \\ 1/3 & 0 & 0 & 0 \end{bmatrix} \quad M_1 = \begin{bmatrix} 0 & 1/2 & 0 & 1/6 \\ 1/2 & 1 & 1/3 & 1/2 \\ 0 & 1/3 & 0 & 0 \\ 1/6 & 1/2 & 0 & 0 \end{bmatrix}$$

$$M_2 = \begin{bmatrix} 0 & 0 & 1/2 & 1/6 \\ 0 & 0 & 1/3 & 0 \\ 1/2 & 1/3 & 1 & 1/2 \\ 1/6 & 0 & 1/2 & 0 \end{bmatrix} \quad M_3 = \begin{bmatrix} 0 & 0 & 0 & 1/3 \\ 0 & 0 & 1/6 & 1/2 \\ 0 & 1/6 & 0 & 1/2 \\ 1/3 & 1/2 & 1/2 & 1 \end{bmatrix}$$

Firstly, it can be seen that we have used the symmetric versions of the matrices as explained previously. Secondly, note that each matrix has the same collection of probabilities, shuffled around. The natural group action on this search space is to consider $\mathcal{X}$ itself as a group under bitwise addition modulo 2. The group table for this is

	00	01	10	11
00	00	01	10	11
01	01	00	11	10
10	10	11	00	01
11	11	10	01	00

The element $\mathbf{0} = (0\,0)$ is the identity in this group. The corresponding permutation matrices are

$$\sigma_0 = \begin{bmatrix} 1 & 0 & 0 & 0 \\ 0 & 1 & 0 & 0 \\ 0 & 0 & 1 & 0 \\ 0 & 0 & 0 & 1 \end{bmatrix} \quad \sigma_1 = \begin{bmatrix} 0 & 1 & 0 & 0 \\ 1 & 0 & 0 & 0 \\ 0 & 0 & 0 & 1 \\ 0 & 0 & 1 & 0 \end{bmatrix}$$

$$\sigma_2 = \begin{bmatrix} 0 & 0 & 1 & 0 \\ 0 & 0 & 0 & 1 \\ 1 & 0 & 0 & 0 \\ 0 & 1 & 0 & 0 \end{bmatrix} \quad \sigma_3 = \begin{bmatrix} 0 & 0 & 0 & 1 \\ 0 & 0 & 1 & 0 \\ 0 & 1 & 0 & 0 \\ 1 & 0 & 0 & 0 \end{bmatrix}$$

The mixing operator $\mathcal{M}$ for this crossover commutes with the group, which means that
$$M_k = \sigma_k M_0 \sigma_k^T$$
for each k. What this means in terms of the effect of crossover on two strings is as follows. Consider the probability that crossing 0 1 with 1 0 will produce 0 0. There is only one way to do it: put the crosspoint in the middle and combine the first element of the first parent with the second element of the second parent. Now consider the probability of crossing 0 0 with 1 1 to produce 0 1. We have an exactly symmetric situation. The two parent strings are complements and we need to combine the first element of one with the second element of the other. It is these symmetries that the group action is describing. From the group table we can see that to relabel 0 1 so it becomes 0 0, we have to 'add' the string 0 1. Applying this action to the whole example shows that they are symmetric:

	example one	group action	example two
first parent	0 1	add 0 1	0 0
second parent	1 0	add 0 1	1 1
desired child	0 0	add 0 1	0 1

It is simple to check that bitwise mutation also commutes with this group.

6.5.1 Schemata

Once we have a group that acts transitively on the search space, and which commutes with mixing, it is interesting to consider the relationship between the detailed structure of the group and the genetic operators. In particular, we are interested in the *subgroup* structure. A subgroup is a subset of the group that is also a group in its own right. To simplify matters we shall only look at the case of fixed-length binary strings, although the results generalise to other search spaces. Consider binary strings of length 5, with bitwise addition modulo 2. Here is an example of a subgroup of this group:

{0 0 0 0 0, 0 0 0 0 1, 0 0 0 1 0, 0 0 0 1 1, 1 0 0 0 0, 1 0 0 0 1, 1 0 0 1 0, 1 0 0 1 1}

Adding together any two strings from this set gives us another member of the set. The identity element $0 = (0 0 0 0 0)$ is also in this set. Note also that this set corresponds to the schema $(*00**)$. In general, any subgroup of $L(\mathcal{X})$ corresponds to a schema. If we now add the element $(0 1 1 0 0)$ to each element in the subgroup, we obtain the following *coset* of the subgroup

{0 1 1 0 0, 0 1 1 0 1, 0 1 1 1 0, 0 1 1 1 1, 1 1 1 0 0, 1 1 1 0 1, 1 1 1 1 0, 1 1 1 1 1},

6.5. REPRESENTATIONAL SYMMETRY

which corresponds to the schema $(*11**)$. In general, if we have a subgroup $A \in L(\mathcal{X})$, then a schema is any set

$$b \circ A(\mathbf{0}) = \{i \in \mathcal{X} : i = (b \circ a)(\mathbf{0}) \text{ for some } a \in A\}$$

for any $b \in L(\mathcal{X})$. We are interested in schemata here because of their intrinsic relationship to the structure of the search space. If these subgroup symmetries are important to us then we might wish to construct a crossover operator that preserves them. Using the terminology introduced by Radcliffe [198], we shall say that crossover *respects* a set $S \subseteq \mathcal{X}$ if, whenever two parents are chosen from S, the offspring is guaranteed to be in S also. We are therefore particularly interested in crossover operators that respect schemata. In the case of binary strings, there is an important class of crossovers that have this property, defined in terms of a binary *mask* according to some probability distribution, as described in Chapter 2. We then assemble the offspring from the parents according to that mask. For example, if we have parents (0 1 1 0 0) and (1 1 1 0 1), the mask (1 1 1 0 0) indicates that we should take the first three bits of the first parent and the last two bits from the second parent to produce (0 1 1 0 1). This string is a member of the schema $(*110*)$, as were both parents: this schema has been respected by the action of crossover. We obtain different crossover operators according to the probability distribution over masks. It is easy to see that all the usual crossovers on binary strings (1X, 2X, UX) can be defined in this way. It can be proved that any crossover defined by masks preserves schemata for binary strings. Moreover, this result can be generalised to a class of search spaces called *structural* spaces. The corresponding crossover operators are called *structural crossovers* and they can be shown to respect the schemata given by the underlying subgroup structure [245].

It should be emphasized that schemata correspond to structural properties of the search space, without reference to the fitness function. There is therefore no guarantee that choosing a crossover that respects them will give a useful operator for any specific search problem. At present there is little theory to relate the structures implicit in a representation to properties of the fitness function. Choices of representation and operators must be made by the practitioner on the basis of whatever is known (or guessed) about the particular problem to be solved. The theory of genetic algorithms cannot at present guide this choice, but simply analyses the connections between the choices of representation and operators. The problem of analysing structures within a particular fitness landscape is addressed in Chapter 9; it is hoped that, eventually, a theory will be developed to connect the representation/operator question with fitness issues. This remains perhaps the single most important research topic in the study of genetic algorithms.

The most obvious attempt to address this issue is of course the Schema Theorem. As we have discussed at length in Chapter 3, this attempts to trace what happens to members of a particular schema from one generation to the next [124]. Although we have just seen that schemata arise naturally in the study of crossover, problems immediately arise in trying to define the action of selection (the operator related to fitness) on a schema. It is usual to define the fitness of a schema to be the average of all members of a population belonging to that schema. However, this gives us a definition that is a function of the particular population (and therefore implicitly a function of time). We cannot, therefore, study what is happening to a particular schema without keeping track of what is happening to the entire population. The Schema Theorem does not buy us anything in terms of theoretical understanding and analysis without a great deal of further work [273, 274, 297].

6.5.2 Walsh transforms

As we shall see in Chapter 9, one approach to the analysis of a fitness landscape is in terms of its Fourier transform. The relationship between the Fourier transform of a fitness function and the ability of operators to search on the corresponding landscape is not well understood. However, this technique does simplify the mixing equations in a remarkable way. In the case of fixed-length binary strings, the Fourier transform is actually the *Walsh transform*.

We have already introduced this transform in Chapter 3; here we use a slightly different version of the matrix $\widetilde{W}$ that represents the Walsh transform. We denote this by

$$W_{i,j} = \frac{1}{\sqrt{n}}(-1)^{i^T j} \quad \text{for } i,j \in \mathcal{X}$$

where $i^T j$ is an inner product.[4] Since W is a real symmetric orthogonal matrix, it is therefore its own inverse, i.e.,

$$W^2 = I.$$

---

[4] The use of the $1/\sqrt{n}$ factor is for reasons of symmetry in defining the inverse transform. In this form the Walsh coefficients differ from the definition in Chapter 3 by a multiplicative constant.

A key property of the Walsh transform is that it simultaneously diagonalizes the permutations matrices σ_k. That is $W\sigma_k W$ is a diagonal matrix for each $k \in \mathcal{X}$. This enables the mixing operator to be simplified:

$$\begin{aligned}\mathcal{M}(Wx)_k &= (Wx)^T M_k Wx \\ &= (Wx)^T \sigma_k M_0 \sigma_k Wx \\ &= x^T W \sigma_k M_0 \sigma_k Wx \\ &= x^T (W\sigma_k W)(WM_0 W)(W\sigma_k W)x\end{aligned}$$

Thus we have the product of the Walsh transform of the mixing matrix with two diagonal matrices. This is a rather efficient way to calculate $\mathcal{M}$. Moreover, it also allows results concerning the inverse of $\mathcal{M}$ and its differential to be obtained [292].

If our search space comprises fixed-length strings in which each position has its own finite alphabet, then a more general transform (the Fourier transform) can be defined which again diagonalizes the associated permutation matrix group. However, it turns out that such spaces are essentially the only ones for which this can be done. The crucial factor is that the underlying groups for such spaces are commutative (i.e., $a \circ b = b \circ a$ for all elements $a, b \in L(\mathcal{X})$). Indeed, any commutative group has a representation of this form (although the crossover operator need not be defined by masks with respect to this structure). However, when the group $L(\mathcal{X})$ is not commutative (for example, the group of permutations in the travelling salesman problem), there is no matrix transform that will diagonalize its matrices.

Given that the Fourier (or Walsh) transform unravels the action of mixing in such an elegant way, and, moreover, that the very same transform can be used to analyse the structure of fitness landscapes, it gives us some hope that this will provide a fruitful avenue for exploring the relationship between particular fitness functions and the genetic operators that act upon them.

6.6 Bibliographic Notes

The study of GAs as dynamical systems has been primarily developed by Michael Vose. His book *The Simple Genetic Algorithm* is a comprehensive description of this theory, applied to search spaces of fixed-length binary strings [293]. This book is essential reading for anyone interested in understanding genetic algorithm theory.

The dynamics of the genetic algorithm with selection and mutation has been extensively analysed by researchers at the Santa Fe Institute, on a range of fitness functions[284]. Their work emphasizes the view of evolution moving along neutral networks in metastable regions and then escaping to new regions. The Z transformation mapping the flow of a selection plus mutation algorithm to that of a pure selection algorithm is also based on their work. Some interesting results for selection and mutation acting on functions of unitation can be found in a paper by Jonathan Rowe [242].

This emergent view of population dynamics has also been propounded by Vose. He has described the long term behaviour of a genetic algorithm as taking place on a graph, in which the vertices correspond to metastable states, and the edges the transitions which can occur between these states. There is, as yet, no general analytical way of constructing this graph, and this remains an important open question.

The analysis of mixing (crossover and mutation) is again based on Vose's presentation. That a mixing matrix might exist for other search spaces was shown by Vose and Darrell Whitley [296]. The more general investigation of the links between group symmetries of the search space and the action of mixing has been developed by Jonathan Rowe, Michael Vose and Alden Wright [244]. The concept of *respect*, which this paper builds on, was introduced by Nick Radcliffe [198, 202].

The relationship between mixing and the Walsh transform was analysed by Vose and Wright for binary strings [292]. The proof that such a transform does not exist for non-commutative groups is in a paper by Rowe, Vose and Wright [245]. Further generalisations can be made—for example, to search spaces with variable-sized structures, as in Genetic Programming (see the work of Riccardo Poli [189] for details). Related work by Chris Stephens has tried to reconcile the dynamical system view of genetic algorithms with the Schema Theorem [273, 274].

Exercises

1. Show that Theorem 6.2 holds for both proportional and rank selection (with ties broken randomly). For arbitrary selection, describe sufficient conditions for this theorem to apply.

2. Write down the operators $\mathcal{F}$ and $\mathcal{U}$ for the genetic algorithm described in Exercise 5 of Chapter 5. Find the fixed points of the infinite population system. Compare your findings with the experimental results.

3. Parameterized uniform crossover is defined in a similar way to uniform crossover, except the probability of a bit being taken from parent 1 is

6.6. BIBLIOGRAPHIC NOTES

 u and the probability that it should come from the second parent is $1 - u$, where u is a parameter of the operator. ('Standard' uniform crossover corresponds to setting $u = 0.5$.) Write down the mixing matrix for this kind of crossover in terms of u, for binary strings of length 2. Remember that the mixing matrix should be symmetric.

4. Write down the mutation matrix U for binary strings of length 2, with a bitwise mutation rate of μ. Calculate the Walsh Transform of the mutation matrix, WUW. What do you notice about this matrix?

5. Write down the mixing matrix M for binary strings of length 2, with uniform crossover. Calculate the Walsh Transform of the mixing matrix, WMW. What do you notice about this matrix?

6. Suppose we have a fitness function f on fixed-length binary strings. Calculate the Walsh Transform of $\mathrm{diag}(f)$, showing that the i,jth component of this matrix is

$$\left(\frac{1}{\sqrt{n}}\right) \widehat{f}_{i \oplus j}$$

where $\widehat{f} = Wf$ is the Walsh Transform of the fitness function. What does this tell you about the problems of handling selection in the Walsh basis?

Chapter 7

Statistical Mechanics Approximations

The equation $\mathcal{G}(p) = \mathcal{M} \circ \mathcal{F}(p)$ that we analysed in the previous chapter describes what happens to a population p in a single generation of a genetic algorithm. It tells us exactly what the probability distribution over all possible next populations is. In the infinite population limit, it tells us the exact trajectory of the population in all subsequent generations. The population p is a vector in n-dimensional space, and so we really have n equations, one for each element of the search space $\mathcal{X}$. In practice, of course, n is extremely large and we have a large number of equations to solve if we want to determine the trajectory of p. This task rapidly becomes impracticable as the size of the search space grows, so it is natural to investigate methods for obtaining an approximation.

7.1 Approximating GA Dynamics

The situation is analogous to one that is often encountered in physics, in which there is a system comprised of many components interacting together. Although we might be able to write down the equation describing each component, there are so many that it is impossible to calculate what happens to the system as a whole. For example, consider a balloon containing some gas. There are so many molecules moving around inside the balloon that it seems hopeless that we shall ever be able to calculate what happens to the balloon as a whole, even if we were given precise information about the state of each molecule. The way to proceed, of course, is to look at average values of the whole system. While we may not be able to predict which molecules strike the surface of the balloon at any one time, we can work out on average how many will do so, and we can summarise this in a single quantity, the pressure of the gas. Similarly, the average velocity of each molecule can be summarised by the temperature of the gas. In this way, we can describe the system as a whole with a couple of parameters, instead of the many required to describe each individual molecule. The branch of physics that deals with such averaging processes is called *statistical mechanics*.

We might therefore hope to describe a genetic algorithm using a similar approach [196, 263]. If we can identify certain average properties of the population, then we might be able to write down equations for these properties, and use these to find out the trajectory of the system as a whole. Of course, in reducing the system to a few average properties we are throwing away a lot of information, and strictly speaking we obtain an approximate result only. The hope is that deviations from this approximation will be relatively rare. In the case of the balloon, it is a possibility that all the molecules in the balloon might all at some time be on one side of the balloon. However, this state of affairs is so unlikely that it seems safe to disregard it. If we assume that our genetic algorithm uses an infinite population, we can similarly disregard fluctuations. As we have seen, in the infinite population limit, the genetic algorithm behaves deterministically. We shall therefore continue to assume that the population is infinite for the moment.

However, if the population is small, then we shall no longer be able to ignore these effects. If the balloon were to contain only two molecules, the chance of both of them being on one side of the balloon is quite high. Similarly, we shall have to consider fluctuations caused by having a finite population. We shall do this by viewing them as *sampling* effects, which cause the population to deviate from the infinite population trajectory.

So how do we go about finding a suitable small set of parameters that (approximately) describe the state of a genetic algorithm? The full system is described by the n variables, $p_0, \ldots, p_{n-1}$, where p_k is the proportion of the population occupied by element k of the search space. We can view these variables as describing a probability distribution over the search space. The mapping $\mathcal{G}$ gives us another such distribution. So we need a good way of approximating distributions.

A well-known method for approximating continuous functions of a real variable is to look at their Fourier series. Any continuous periodic function can be expressed as an infinite sum of sine and cosine functions. For example, consider the function $y = x^2$ on the interval $-1 < x < 1$. We can calculate

7.1. APPROXIMATING GA DYNAMICS

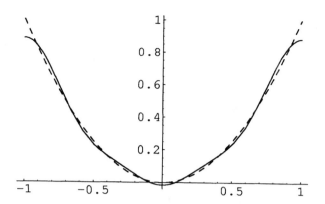

Figure 7.1: The curve $y = x^2$ (dashed line) and its approximation by its Fourier series truncated after $k = 3$.

the Fourier series of this function to be

$$y = \frac{1}{3} + \frac{4}{\pi^2} \sum_{k=1}^{\infty} \frac{(-1)^k \cos k\pi x}{k^2}$$

or

$$y = \frac{1}{3} - \frac{4}{\pi^2} \cos \pi x + \frac{1}{\pi^2} \cos 2\pi x - \frac{4}{9\pi^2} \cos 3\pi x + \dots$$

Clearly, as the series is expanded, the terms correspond to higher and higher frequencies, and have smaller and smaller coefficients. We can think of the first few terms as providing an approximation to the curve $y = x^2$, and the later terms as giving small corrections (see Figure 7.1).

The distributions that we wish to approximate are not continuous, but discrete. However, we shall see that a similar approach can be used in this case. We shall show that the population equation can be written as an infinite series with ever decreasing coefficients, and that we can therefore approximate the characteristics of the infinite population by truncating this series. It may seem a backward step to express a set of n variables using an infinite set of coefficients. The trick is, of course, that we shall truncate this series to obtain an approximation using far fewer than n terms. These leading coefficients will be our summary statistics with which we shall track the trajectory of the population.

7.2 Generating Functions

Suppose we have a random variable X that can take non-negative integer values. For each $k \geq 0$ there is a certain probability that X takes the value k. The probability distribution for X can be described by its *probability generating function*

$$G(x) = \sum_{k \geq 0} \Pr[X = k] x^k.$$

For example, if we have a population vector $\boldsymbol{p} = (p_0, p_1, p_2, p_3)$ then the corresponding probability generating function is

$$G(x) = p_0 + p_1 x + p_2 x^2 + p_3 x^3.$$

This polynomial contains complete information about the population $\boldsymbol{p}$. We can therefore work just as well with these functions as with the population vector itself. It is easy to see that for any probability generating function

$$G(1) = 1$$

and that its derivative is

$$G'(x) = \sum_{k > 0} k \Pr[X = k] x^{k-1},$$

so that

$$G'(1) = E[X].$$

To proceed to our goal of expressing the population distribution as an infinite series, we use the substitution of variables

$$x = e^z$$

which gives us the *characteristic function* of the distribution

$$\varphi(z) = \sum_{k \geq 0} \Pr[X = k] e^{kz}.$$

If we expand the characteristic function as a Taylor series in z we obtain

$$\varphi(z) = 1 + \frac{\mu_1}{1!} z + \frac{\mu_2}{2!} z^2 + \frac{\mu_3}{3!} z^3 + \dots,$$

where the numbers μ_k are called the *moments* of the distribution, being given by

$$\mu_k = E[X^k]$$

If we are given $\varphi(z)$ we can find the moments by differentiating an appropriate number of times and setting $z = 0$:

$$\mu_k = \varphi^{(k)}(0).$$

7.2.1 Cumulants

Is this the infinite series we are looking for, by which we can approximate our population p? Unfortunately not! The problem is that the moments grow larger and larger, not smaller and smaller, since $\mu_k = E[X^k]$. Instead we look at a related function, called the *cumulant generating function* (*cgf*), and its Taylor series:

$$H(z) = \log \varphi(z) = \frac{\kappa_1}{1!}z + \frac{\kappa_2}{2!}z^2 + \frac{\kappa_3}{3!}z^3 + \ldots .$$

The coefficients κ_k are called the *cumulants* of the distribution. There is a series of formulae for converting between moments and cumulants, which starts like this:

$$\kappa_1 = \mu_1$$
$$\kappa_2 = \mu_2 - \mu_1^2$$
$$\kappa_3 = \mu_3 - 3\mu_1\mu_2 + 2\mu_1^2$$

We see that the first cumulant κ_1 is the mean of the distribution, and the second κ_2 is the variance. The Gaussian (or Normal) distribution is completely defined by its mean and variance and all the higher cumulants are zero. Any distribution with very small values for the higher cumulants will look rather like a Gaussian. This, then, is the infinite series that we shall use to approximate populations. As long as populations are distributed approximately normally, we shall be able to truncate the series expansion and use just the first few cumulants.

Although the definition of cumulants looks a lot uglier than that for moments, they are in fact much simpler to use. The main reason for this is that they have the property of linearity: if we know the *cgf* for each of two independent random variables X and Y, then

$$H_{X+Y}(z) = H_X(z) + H_Y(z)$$

so that the cumulants for the random variable $X + Y$ can be found simply as a sum

$$\kappa_j^{X+Y} = \kappa_j^X + \kappa_j^Y,$$

a property that is true for moments only in the case of μ_1. This property of cumulants will prove very useful later.[1]

---
[1] More information on probability generating functions and cumulants can be found in Chapter 8 of [105].

178   CHAPTER 7. STATISTICAL MECHANICS APPROXIMATIONS

The first problem we face in using this approach is to decide what the random variable should be. In the definition of the probability generating function given above, we took it that the variable X was ranging over the search space $\mathcal{X}$. However, when thinking about the effects of selection, it is more convenient to think of X as taking on the different possible *fitness values*. In other words, X must range over

$$f(\mathcal{X}) = \{x \in \mathbb{R}^+ : x = f(k) \text{ for some } k \in \mathcal{X}\}$$

Although this choice is convenient for selection, it is not necessarily useful for handling mutation and crossover. These operators act on the structure of the elements of the search space, independently of the fitness. This dilemma is really the same problem that we came across in the last chapter. The effects of mutation and crossover are *syntactic* (they affect the structure of elements of $\mathcal{X}$) while the effect of selection is *semantic* (it depends on the 'meaning' of an element, as described by the fitness function). To make life simple we shall concentrate on the one particular case where there is no (or little) distinction between the two: the *Onemax* problem. Here, our search space is the set of binary strings of length ℓ and the fitness function is simply the number of ones in the string. With this function, all we need to be concerned about is the number of ones within a string: the *unitation class* of the string. We shall therefore think of our random variable X as ranging over the set of possible unitation classes $\{0, 1, 2, \ldots, \ell\}$. (Further examples of functions defined over unitation classes can be found in Appendix A.)

Here is an example. Suppose our string length is $\ell = 5$ and we have a population of size 20. If we generate an initial population at random, we might obtain the following:

```
11000   00111   01100   10111
11100   11110   01110   00101
01101   01010   01101   01110
10100   01001   11010   00110
00001   01000   11000   11000
```

If we view this population as defining a distribution over the unitation classes, we obtain

$$\begin{aligned}
\mathbf{Pr}[X = 0] &= 0 \\
\mathbf{Pr}[X = 1] &= 0.1 \\
\mathbf{Pr}[X = 2] &= 0.45 \\
\mathbf{Pr}[X = 3] &= 0.35
\end{aligned}$$

7.3. SELECTION

$$\Pr[X = 4] = 0.1$$
$$\Pr[X = 5] = 0$$

The probability generating function for X is the polynomial

$$G(x) = 0.1x + 0.45x^2 + 0.35x^3 + 0.1x^4.$$

Making the substitution $x = e^z$ gives us the characteristic function

$$\varphi(z) = 0.1e^z + 0.45e^{2z} + 0.35e^{3z} + 0.1e^{4z}.$$

We can expand this as a Taylor series to generate the moments of the distribution

$$\begin{aligned}\varphi(z) &= 1 + 2.45z + 3.325z^2 + 3.258z^3 + 2.552z^4 + \ldots \\ &= 1 + \frac{2.45}{1!}z + \frac{6.65}{2!}z^2 + \frac{19.548}{3!}z^3 + \frac{61.248}{4!}z^4 + \ldots\end{aligned}$$

from which we can read off the moments $\mu_1 = 2.45, \mu_2 = 6.65$ and so on. As can be seen, the moments grow larger and larger. Taking the natural logarithm of φ gives us the cumulant generating function

$$H(z) = \log \varphi(z) = \log(0.1e^z + 0.45e^{2z} + 0.35e^{3z} + 0.1e^{4z} + \ldots).$$

Expanding this as a Taylor series gives us the cumulants

$$\begin{aligned}H(z) &= 2.45z + 0.3238z^2 + 0.0141z^3 - 0.0078z^4 + \ldots \\ &= \frac{2.45}{1!}z + \frac{0.6475}{2!}z^2 + \frac{0.08475}{3!}z^3 - \frac{0.188}{4!}z^4 + \ldots\end{aligned}$$

We see immediately that the mean is 2.45 and the variance 0.6475. A smoothed version of this distribution is nearly Gaussian (see Figure 7.2), so we can approximate the population by truncating this series expansion. We might therefore only use the first three or four cumulants as our 'macroscopic' variables and consider what happens to them under selection, mutation and crossover.

7.3 Selection

Suppose we have a population p with corresponding characteristic function

$$\varphi_p(z) = \sum_{k \geq 0} \Pr[X = k \text{ in } p]e^{kz}$$

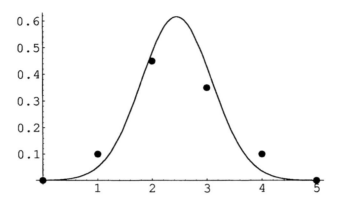

Figure 7.2: The distribution of X over unitation classes $0, \ldots, 5$. The Gaussian distribution with the same mean and variance is shown as a solid line.

where our random variable X ranges over possible fitness values. That is, we can write
$$\varphi_{\boldsymbol{p}}(z) = \sum_{k \geq 0} p_k e^{kz}$$
where p_k is the proportion of the population with fitness k. We want to know the effect of applying proportional selection to $\boldsymbol{p}$. In the infinite population limit, this gives us a new population vector $\boldsymbol{q} = \mathcal{F}(\boldsymbol{p})$. The characteristic function for the new population is
$$\varphi_{\boldsymbol{q}}(z) = \sum_{k \geq 0} \Pr[X = k \text{ in } \boldsymbol{q}] e^{kz}$$
The probability of a fitness value k appearing in the new population is
$$q_k = \frac{k p_k}{\mu_1(\boldsymbol{p})}$$
where $\mu_1(\boldsymbol{p})$ is the average fitness in population $\boldsymbol{p}$. This is just the equation for selection that we worked out in the previous chapter. We therefore have a characteristic function
$$\begin{aligned}
\varphi_{\boldsymbol{q}}(z) &= \sum_{k \geq 0} \frac{k p_k}{\mu_1(\boldsymbol{p})} e^{kz} \\
&= \frac{1}{\mu_1(\boldsymbol{p})} \sum_{k \geq 0} k p_k e^{kz} \\
&= \frac{1}{\mu_1(\boldsymbol{p})} \sum_{k \geq 0} p_k \frac{d}{dz} e^{kz}
\end{aligned}$$

7.3. SELECTION

$$\begin{aligned}
&= \left(\frac{1}{\mu_1(\boldsymbol{p})}\right) \frac{d}{dz} \sum_{k \geq 0} p_k e^{kz} \\
&= \left(\frac{1}{\mu_1(\boldsymbol{p})}\right) \frac{d}{dz} \varphi_{\boldsymbol{p}}(z)
\end{aligned}$$

Thus, we can find the new characteristic function by differentiating the old one and dividing by the mean fitness. This makes it particularly easy to obtain the new moments. If we write the Taylor expansion of $\varphi_{\boldsymbol{p}}(z)$ as

$$\varphi_{\boldsymbol{p}}(z) = 1 + \frac{\mu_1}{1!} z + \frac{\mu_2}{2!} z^2 + \frac{\mu_3}{3!} z^3 + \ldots$$

then differentiating with respect to z gives us

$$\frac{d}{dz} \varphi_{\boldsymbol{p}}(z) = \mu_1 + \frac{\mu_2}{1!} z + \frac{\mu_3}{2!} z^2 + \ldots$$

Dividing through by μ_1 then gives us the new characteristic function

$$\varphi_{\boldsymbol{q}}(z) = 1 + \frac{\mu_2}{\mu_1 1!} z + \frac{\mu_3}{\mu_1 2!} z^2 + \ldots$$

so we can read off the moments of the new population $\boldsymbol{q} = \mathcal{F}(\boldsymbol{p})$:

$$\begin{aligned}
\mu_1(\boldsymbol{q}) &= \frac{\mu_2(\boldsymbol{p})}{\mu_1(\boldsymbol{p})} \\
\mu_2(\boldsymbol{q}) &= \frac{\mu_3(\boldsymbol{p})}{\mu_1(\boldsymbol{p})}
\end{aligned}$$

and, in general

$$\mu_k(\boldsymbol{q}) = \frac{\mu_{k+1}(\boldsymbol{p})}{\mu_1(\boldsymbol{p})}$$

However, we really need to use cumulants rather than moments. This is a pity in some ways because the moment equations for proportional selection are so straightforward. However, we can use these formulae to convert between moments and cumulants, and after a little rearranging we obtain

$$\begin{aligned}
\kappa_1(\mathcal{F}(\boldsymbol{p})) &= \kappa_1(\boldsymbol{p}) + \left(\frac{\kappa_2(\boldsymbol{p})}{\kappa_1(\boldsymbol{p})}\right) \\
\kappa_2(\mathcal{F}(\boldsymbol{p})) &= \kappa_2(\boldsymbol{p}) - \left(\frac{\kappa_2(\boldsymbol{p})}{\kappa_1(\boldsymbol{p})}\right)^2 + \left(\frac{\kappa_3(\boldsymbol{p})}{\kappa_1(\boldsymbol{p})}\right) \\
\kappa_3(\mathcal{F}(\boldsymbol{p})) &= \kappa_3(\boldsymbol{p}) + 2\left(\frac{\kappa_2(\boldsymbol{p})}{\kappa_1(\boldsymbol{p})}\right)^3 - \left(\frac{3\kappa_2(\boldsymbol{p})\kappa_3(\boldsymbol{p})}{\kappa_1(\boldsymbol{p})}\right) + \left(\frac{\kappa_4(\boldsymbol{p})}{\kappa_1(\boldsymbol{p})}\right)
\end{aligned}$$

and so on. Clearly, it will become rather ugly for the higher cumulants, but fortunately it is these that we shall truncate!

Note that we have in fact derived these equations for proportional selection for *any* fitness function. Nowhere did we need to restrict ourselves to functions of unitation. We can immediately gain some qualitative insight into the effects of selection. The first equation tells us that the mean will increase by an amount proportional to the variance, and inversely proportional to the mean of the fitness. So the greater the spread of values in our current population, the greater the increase in mean fitness. This relates quite closely to previously known results in population genetics concerning the *response to selection*, as discussed, for example in the context of the Breeder GA [175].

The second equation tells us what happens to the variance. If the third cumulant is relatively small, it is clear that the variance decreases by an amount that is roughly the square of the ratio of the mean and the variance. We should expect the variance to drop quickly while κ_3 is small. This third cumulant is related to the *skewness* of the distribution. This is a measure of how much of the distribution is on one side of the mean. Its exact behaviour is hard to predict however. In particular, as the population converges around the maximum the distribution may become very skewed, which will seriously compromise our approximations. If the distribution becomes one-sided, we shall have a large number of copies of the maximum fitness value and fewer copies of smaller values, but no copies at all of higher values. The approximation will break down at this point. However, we hope that the *transient* trajectory of the population up to this point will be modelled fairly accurately.

Suppose we assume that the cumulants $\kappa_4, \kappa_5, \ldots$ are zero, which gives us an approximation to the dynamics of selection using just three variables. How well does this perform in modelling the transient stage of evolution on the *Onemax* problem? Firstly we need some initial values for our variables. If the strings are of length ℓ and are generated randomly, the number of ones is given by a binomial distribution of ℓ trials with probability $1/2$. This gives us the following initial conditions:

$$\kappa_1(\boldsymbol{p}(0)) = \frac{\ell}{2}$$
$$\kappa_2(\boldsymbol{p}(0)) = \frac{\ell}{4}$$
$$\kappa_3(\boldsymbol{p}(0)) = 0$$

We now iterate our equations to obtain predictions for the first few generations. The graphs in Figure 7.3 illustrates what happens when $\ell = 10$. The clear prediction is that the mean should increase, the variance decrease and the skewness quickly become zero. It is this last prediction which is almost certainly incorrect as the population approaches the maximum, and this will create errors in the exact predictions of the other two values. The average over 200 runs of an actual GA is also illustrated in this figure. As can be seen, the prediction for the mean fitness is quite good for the first few generations (the transient phase). As the population begins to converge, however, the approximation becomes worse and worse. The variance does decrease, but much faster than predicted. The real problem is κ_3, however, where the prediction is frankly a disaster, except that at least its value does stay relatively small.

We have managed to predict approximately the transient values of the mean for the *Onemax* problem under proportional selection, using three variables only, but the predictions for the other cumulants are not so good. In particular, the whole approximation breaks down close to convergence. As Figure 7.4 shows, this is due to the population becoming more and more lop-sided as it approaches the maximum value. Happily, this problem will be alleviated when we introduce mutation and crossover because, amongst other things, these operators have the effect of maintaining the population distribution closer to a bell-shaped (Gaussian) curve throughout the evolutionary process.

7.4 Mutation

We now add bitwise mutation to our genetic algorithm, following the development of [196]. As remarked above, this helps to keep the population closer to a Gaussian distribution throughout the process. This can be seen in a typical run on *Onemax* ($\ell = 10$), as shown in Figure 7.5. It is evident that, even when the process has converged, we still obtain a fairly symmetric population. Thus, the higher cumulants stay small and we should obtain a much better approximation than we did with selection only.

To incorporate mutation into our model we must calculate the effects this operator has on the cumulants of a population, and then combine these effects with those we worked out for proportional selection. Restricting our attention to *Onemax* makes this task fairly easy. Recall that we are considering a random variable X that ranges over the possible fitness values. We can think of the value of each bit position as being a random variable that can take on the values 0 or 1. The fitness is then given by the sum of

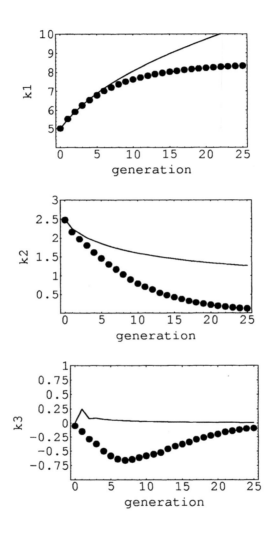

Figure 7.3: The predicted and actual values for κ_1 (top), κ_2 (centre) and κ_3 (bottom) for proportional selection for *Onemax* with $\ell = 10$. The theoretical predictions are shown as solid lines.

7.4. MUTATION

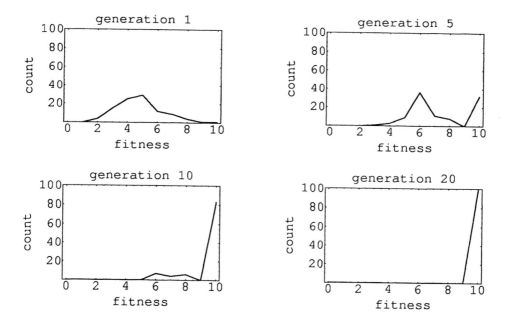

Figure 7.4: The population distribution for an actual run of a GA using selection only (population size is 100). Initially the distribution is fairly symmetric but after a few generations it becomes totally one-sided.

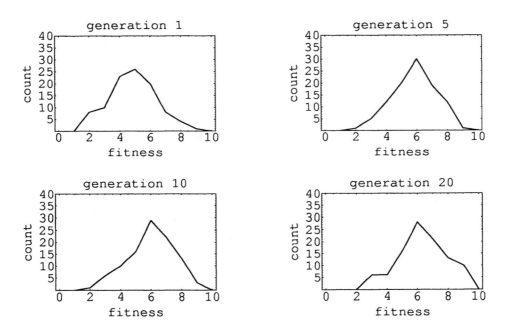

Figure 7.5: The population distribution for a typical run with mutation (with probability 0.1) and selection. Population size is 100. The population distribution now stays fairly symmetric throughout the run.

7.4. MUTATION

these variables. Denoting the value of bit j by X_j, our fitness variable is

$$X = \sum_j X_j$$

If we can work out the cumulants for each X_j then we can calculate the cumulants for X simply by addition. (This follows from the helpful linear property that we have already mentioned.) So if we denote the cumulants for X_j by $\kappa_1^{(j)}, \kappa_2^{(j)}, \ldots$, we have

$$\kappa_k = \sum_j \kappa_k^{(j)}.$$

Note that we have to assume that the values of each bit position are independent of each other. This assumption is satisfied for *Onemax*, but its validity needs re-evaluating for any problem with non-linear interactions between the bits (epistasis).

If we consider a single bit position as a random variable X_j, we can first calculate the moments and then convert them to cumulants as before. Recall that the moments of the distribution are given by

$$\mu_k^{(j)} = E[X_j^k].$$

In this case we have a remarkable property: because X_j can only take the values 0 or 1 it follows that

$$X_j^k = X_j$$

for all values of k, and thus moments of all orders are equal! This simplifies things greatly: we only need work out the effect of mutation on the mean of X_j and we automatically have the result for all moments. Suppose that we have a randomly distributed population and consider the effect of mutation on the values in a single bit position. There are two ways in which a bit can end up as 1 after mutation: it can start as 0 and then mutate, or it can start as 1 and fail to mutate. Similarly, there are two ways that the bit can end up as 1. If the probability of mutating[2] is u, we can compute the expected value of the bit after mutation:

$$\begin{aligned} E[X_j'] &= \sum_{i=0,1} i \mathbf{Pr}[X_j' = i] \\ &= \mathbf{Pr}[X_j' = 1] \end{aligned}$$

---

[2] Note that we can't use our usual notation of μ for mutation rate in this chapter as it conflicts with the conventional notation for moments.

$$\begin{aligned}
&= u\mathbf{Pr}[X_j = 0] + (1-u)\mathbf{Pr}[X_j = 1] \\
&= u(1 - \mathbf{Pr}[X_j = 1]) + (1-u)\mathbf{Pr}[X_j = 1] \\
&= u + (1 - 2u)\mathbf{Pr}[X_j = 1]
\end{aligned}$$

(Note how the case $i = 0$ disappears from the sum.) Now we want to express the mean after mutation in terms of the mean before mutation. This is simply

$$E[X_j] = \mathbf{Pr}[X_j = 1]$$

and so we can write

$$E[X_j'] = u + (1 - 2u)E[X_j].$$

Since all the moments are equal, we can immediately write down

$$\mu_k^{(j)}(\boldsymbol{q}) = u + (1-2u)\mu_k^{(j)}(\boldsymbol{p}) = u + (1-2u)\mu_1^{(j)}(\boldsymbol{p}) \quad \forall k,$$

where $\boldsymbol{q} = \mathcal{U}(\boldsymbol{p})$.

We now need to convert from moments to cumulants, so that we can sum them. The first one is easy, since it is just the mean. We therefore have

$$\kappa_1^{(j)}(\boldsymbol{q}) = u + (1 - 2u)\kappa_1^{(j)}(\boldsymbol{p})$$

Thus the mean fitness of the string after mutation is

$$\begin{aligned}
\kappa_1(\boldsymbol{q}) &= \sum_j \kappa_1^{(j)}(\boldsymbol{q}) \\
&= \sum_j \left(u + (1-2u)\kappa_1^{(j)}(\boldsymbol{p}) \right) \\
&= u\ell + (1-2u)\kappa_1(\boldsymbol{p})
\end{aligned}$$

The second cumulant is, in general, given by the formula

$$\kappa_2 = \mu_2 - \mu_1^2,$$

so after mutation, we have

$$\begin{aligned}
\kappa_2^{(j)}(\boldsymbol{q}) &= u + (1-2u)\mu_1^{(j)}(\boldsymbol{p}) - [u + (1-2u)\mu_1^{(j)}(\boldsymbol{p})]^2 \\
&= u(1-u) + (1-2u)^2(\kappa_1^{(j)}(\boldsymbol{p}) - [\kappa_1^{(j)}(\boldsymbol{p})]^2) \\
&= u(1-u) + (1-2u)^2 \kappa_2^{(j)}(\boldsymbol{p})
\end{aligned}$$

7.4. MUTATION

(The last line follows from the fact that $\kappa_1^{(j)} = \mu_1^{(j)} = \mu_2^{(j)}$.) Summing over all the bits gives us the variance of the fitness after mutation

$$\kappa_2(q) = u(1-u)\ell + (1-2u)^2 \kappa_2(p).$$

We can similarly (if more painfully) derive a formula for the third cumulant

$$\kappa_3(q) = u(1-u)(1-2u)(\ell - 2\kappa_1(p)) + (1-2u)^3 \kappa_3(p).$$

Before we go any further, we should ask what these equations tell us about the effect of mutation. We can rearrange the equation for the mean fitness as

$$\kappa_1(q) = \kappa_1(p) + 2u\left(\frac{\ell}{2} - \kappa_1(p)\right)$$

This indicates that the average unitation of a string will be moved towards the value $\ell/2$. In the case of *Onemax*, of course, the unitation is the same as fitness, but this observation holds whatever the fitness function is, if we just think of mutation acting on the unitation.

Now we can put the mutation equations together with those for fitness-proportional selection. This will give us a three parameter approximate model for a more realistic genetic algorithm. Assuming once again that all cumulants higher than the third are zero, we obtain

$$\kappa_1(\mathcal{U} \circ \mathcal{F}(p)) = u\ell + (1-2u)\left(\kappa_1(p) + \frac{\kappa_2(p)}{\kappa_1(p)}\right)$$

$$\kappa_2(\mathcal{U} \circ \mathcal{F}(p)) = u(1-u)\ell + (1-2u)^2 \left(\kappa_2(p) - \left(\frac{\kappa_2(p)}{\kappa_1(p)}\right)^2 + \frac{\kappa_3(p)}{\kappa_1(p)}\right)$$

$$\kappa_3(\mathcal{U} \circ \mathcal{F}(p)) = u(1-u)(1-2u)\left(\ell - 2\kappa_1(p) - 2\frac{\kappa_2(p)}{\kappa_1(p)}\right)$$

$$+ (1-2u)^3 \left(\kappa_3(p) + 2\left(\frac{\kappa_2(p)}{\kappa_1(p)}\right)^3 - 3\frac{\kappa_2(p)\kappa_3(p)}{\kappa_1(p)}\right)$$

We can now iterate these equations to obtain predictions for the evolution of our GA. Figure 7.6 shows the predicted values for *Onemax* with $\ell = 10$, together with experimental results averaged over 200 runs. This time we see that the prediction for the mean fitness is excellent throughout. The variance is also fairly well predicted, although it is slightly over-estimated. The third cumulant is not so good for the first generation, but settles to a value which is again slightly over-estimated. These over-estimates arise from the fact that we are using a finite population of size 100. This effect will be accounted for in the following section, but even as it stands, we have a fairly impressive set of predictions considering our model uses just three variables.

7.5 Finite Population Effects

So far we have been treating the fitness of a population as a random variable with some probability distribution. We have estimated the cumulants of that distribution, and studied the way they change over time. However, in an actual run of a genetic algorithm, we have a finite population, so that we are effectively sampling our fitness probability distribution N times, where N is the population size. To account for finite population effects we need to address the following question: given a probability distribution, with a certain mean, variance and higher cumulants, what will be the expected value of the mean, variance and higher cumulants of a finite sample from this distribution? Note what we are saying here: given that we are taking a finite sample, the mean of that sample is now itself a random variable! We are interested in the expected value of that variable, as we are for all the cumulants of a finite sample.

Naively, one might assume that the expected values of the cumulants of a finite sample will be that same as the cumulants of the distribution from which the sample is taken. However, this is not the case in general; in fact, the size of the sample affects these values. For the first three cumulants, standard results in statistics tell us that the expected values are

$$E[\kappa_1] = \kappa_1$$
$$E[\kappa_2] = \left(1 - \frac{1}{N}\right)\kappa_2$$
$$E[\kappa_3] = \left(1 - \frac{1}{N}\right)\left(1 - \frac{2}{N}\right)\kappa_3$$

where the terms on the left-hand sides represent the expected values of the cumulants of a sample of size N. Evidently, only the mean is unaffected by sample size. The use of sample statistics to take account of population size effects is one of the main advantages of the statistical mechanics approach [196, 263].

We can now modify our equations for selection and mutation to account for population size by multiplying by the appropriate factors. To illustrate the effect of this, we shall consider a GA with a rather small population size, $N = 20$. The predictions, together with some experimental results,

7.5. FINITE POPULATION EFFECTS

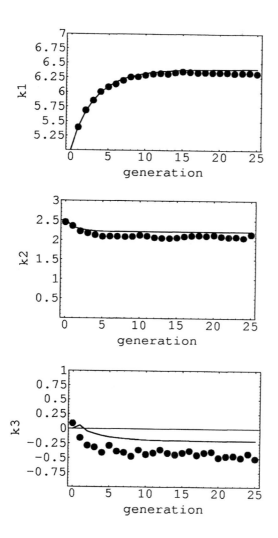

Figure 7.6: The predicted and actual values for κ_1 (top), κ_2 (centre) and κ_3 (bottom) for proportional selection and mutation for *Onemax* with $\ell = 10$. The theoretical predictions are shown as solid lines.

are shown in Figure 7.7 (results are averaged over 200 runs). The original predictions from the previous section are shown as dashed lines. The predictions that include finite population corrections are given as solid lines. In the case of the mean and variance, the new predictions work extremely well. The results are not so good for the third cumulant. This is not entirely surprising, since the effect of selection on the third cumulant depends on the value of the fourth cumulant, which we are ignoring. We would find a better result were we to include it as a fourth variable in our model.

It is also clear that the experimental results show a higher level of fluctuation from generation to generation than we had with a population size of 100. This is a second-order effect of finite populations. In the same way that we have been considering the expected value of the cumulants for a finite sample, we could also examine the variance of these random variables. This would give us a measure of the amount of fluctuation to expect. For example, it is known that the variance of κ_2 is roughly inversely proportional to the sample size N.

7.6 Crossover

When it comes to modelling the effects of crossover, we immediately encounter a problem. The trouble is that two strings with the same fitness can have very different effects when crossed over. For example, suppose we have strings of length 6. The string (0 0 0 1 1 1) when crossed with itself will always remain unchanged (under all 'normal' kinds of crossover). However, if we cross it with the string (1 1 1 0 0 0), which in the case of *Onemax* has the same fitness, we could end up with a variety of different strings. In fact, for the case of complementary strings like these we could generate any string in the Universe with UX. This means that we can't simply ask for the probability of two strings with a given fitness producing a third with some fitness. We also have to take into account the distribution of ones in the strings. The effect of this is that the bits can no longer be considered to be independent variables as they were under mutation, so we can't proceed by finding the sum of the cumulant values for the individual bits.

To model crossover properly requires that we keep track of the correlations between all the bit variables. We also need to consider how that correlation is itself affected by the genetic operators. Pursuing this course is possible, but complex (see, for example, the development in [196]). We shall take the simpler option of making an approximation. Consider what would happen to a population if we just kept on applying crossover again

7.6. CROSSOVER

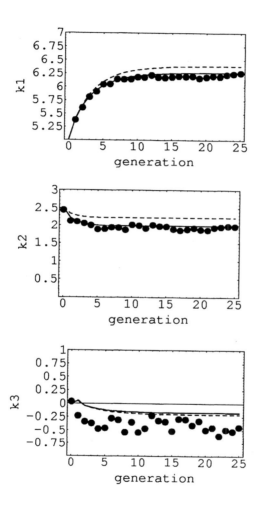

Figure 7.7: The effects of having a small population ($N = 20$). The original predictions are shown as dashed lines. The solid lines include finite population correction factors.

and again. A well-known result from theoretical biology (Geiringer's Theorem) tells us that eventually the values in any given bit position become completely scrambled [90]. More precisely, if there are x_j ones in position j, then after repeated application of crossover, the probability that any string in the population has a one in that position is simply x_j/N, where N is the population size. Furthermore, this is true for each bit position independently. This distribution is known as *Robbins' proportions*, and when the population is distributed like this, we can indeed treat the bit variables independently. How does this help us with modelling a single application of crossover? Well, it is also known that under repeated crossover, the convergence to Robbins' proportions is very fast, although the rate is different for different forms of crossover. So it seems not unreasonable to use this distribution as an approximation even after a single round of crossover. Of course, this can go seriously wrong if we happened to start with a pathological initial population—for example, suppose we always initialized the strings by putting all the ones to the left. But provided that we start with a genuinely random distribution of bits, there is a good chance our approximation won't be too bad.[3]

We can now calculate the effect of crossover on the cumulants of a population using this approximation. A little thought is enough to convince us that crossover does not change the mean fitness at all for *Onemax*. Crossover never creates more ones or zeros: it merely shuffles the bits around. If $q = \mathcal{C}(p)$ we can write down at once that

$$\kappa_1(q) = \kappa_1(p)$$

for crossover. For the variance, recall that we are assuming Robbins' proportions after crossover, so that the bit values are independent. Thus the probability of any one bit in a string being a one is κ_1/ℓ, and we can view each string as being ℓ independent Bernoulli trials; i.e., the fitness of a string is distributed binomially with ℓ trials and probability κ_1/ℓ of having a one in each trial. The variance of such a distribution is given by

$$\begin{aligned}\kappa_2(q) &= \ell\left(\frac{\kappa_1(q)}{\ell}\right)\left(1 - \frac{\kappa_1(q)}{\ell}\right) \\ &= \kappa_1(q)\left(1 - \frac{\kappa_1(q)}{\ell}\right)\end{aligned}$$

---

[3]Incidentally, for the type of 'gene pool' crossover [176] that works on the whole population, rather than on pairs of strings, this isn't an approximation—it's exact!

7.6. CROSSOVER

We can similarly find the third cumulant simply by taking the third cumulant of the binomial distribution and substituting in our Robbins' proportion

$$\kappa_3(q) = \kappa_1(q) \left(1 - \frac{\kappa_1(q)}{\ell}\right) \left(1 - \frac{2\kappa_1(q)}{\ell}\right)$$

We can now incorporate all these results into our previous equations for the effects of mutation and selection and we have a three parameter model for the full genetic algorithm on *Onemax*:

$$\kappa_1(\mathcal{G}(p)) = u\ell + (1 - 2u)\left(\kappa_1(p) + \frac{\kappa_2(p)}{\kappa_1(p)}\right)$$

$$\kappa_2(\mathcal{G}(p)) = \kappa_1(\mathcal{G}(p))\left(1 - \frac{\kappa_1(\mathcal{G}(p))}{\ell}\right)$$

$$\kappa_3(\mathcal{G}(p)) = \kappa_1(\mathcal{G}(p))\left(1 - \frac{\kappa_1(\mathcal{G}(p))}{\ell}\right)\left(1 - \frac{2\kappa_1(\mathcal{G}(p))}{\ell}\right)$$

where $\mathcal{G} = \mathcal{C} \circ \mathcal{U} \circ \mathcal{F}$. Note that our assumption about crossover means that its effect on higher cumulants completely dominates the effects of selection and mutation. The results of iterating these equations is shown in Figure 7.8 along with experimental results (averaged over 200 runs). We see a good level of agreement between theory and practice, which indicates that our simplifying assumption is reasonable. Once again, though, there is a slight over-estimation of the magnitude of the cumulants, which may be due to finite population effects.

On examining the equations more closely, we see that both κ_2 and κ_3 are given solely in terms of κ_1. Thus, we can reduce our set of equations to a single one, giving us the trajectory of the mean fitness:

$$\kappa_1(\mathcal{G}(p)) = u\ell + (1 - 2u) + (1 - 2u)\left(1 - \frac{1}{\ell}\right)\kappa_1(p)$$

This difference equation can be solved to give a closed form equation for the mean—although a rather messy one. Of more interest is the fixed point of the equation, which can be directly calculated as

$$\kappa_1 = \frac{\ell(u\ell + 1 - 2u)}{2u\ell + 1 - 2u}$$

which is the limit value of the mean as $t \to \infty$. This value is shown for different mutation rates in Figure 7.9. It is clear that increasing mutation pushes the fixed point away from the optimum. However, we should be wary of drawing general conclusions from these results, based as they are on such an over-simplification of the effects of crossover for one toy problem!

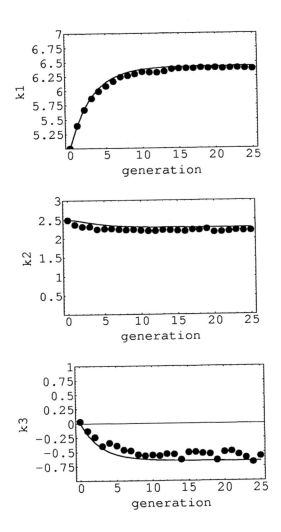

Figure 7.8: The predicted and actual values for κ_1 (top), κ_2 (centre) and κ_3 (bottom) for proportional selection, mutation and uniform crossover for *Onemax* with $\ell = 10$. The theoretical predictions are shown as solid lines.

7.7. BIBLIOGRAPHIC NOTES

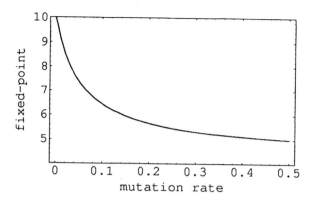

Figure 7.9: The limiting value of the mean fitness, under proportional selection, mutation and crossover, for different mutation rates in the case $\ell = 10$.

7.7 Bibliographic Notes

The statistical mechanics approach to approximating genetic algorithm behaviour was developed by Jonathan Shapiro, Magnus Rattray, Adam Prügel-Bennett, and Alex Rogers. One of the advantages of the approach is that it is possible to model a variety of selection methods, including tournament, rank and Boltzmann selection [23, 196, 263]. Adapting the method for fitness functions other than *Onemax* is difficult. Successes to date include interacting spin systems [195], in which the fitness contribution of a bit depends on the value of neighbouring bits, and the 'basin with a barrier' problem [264]. Spin systems have also been analysed by Bart Naudts [179] from a fitness landscape viewpoint. The 'basin with a barrier' problem is illustrated in Figure 7.10. It is like the *Onemax* problem but with a gap to isolate the optimum.

It can be shown that using crossover enables the population to make the jump across the barrier far more efficiently than mutation alone can manage. A rigorous analysis of running-time complexity on similar fitness functions, by Thomas Jansen and Ingo Wegener [133] has confirmed theoretically the results of [264]. The perceptron [210] and subset-sum problems [209] have also been modelled successfully, and a large amount of work has been done in modelling the effects of crossover rather more accurately than we have done in this chapter. Stefan Bornholdt [27] has applied the same methodology to the analysis of GA behaviour on NK-landscapes (described in Appendix A).

It should be noted that in a number of the papers describing this approach, statistical mechanics terminology is normally used. This can be a

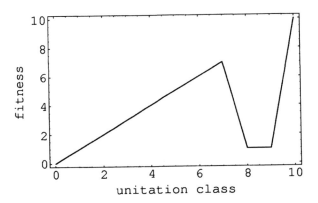

Figure 7.10: The 'basin with a barrier' problem.

little confusing to readers with no physics background, so a short explanation is called for. In most of these papers, what we have been calling *bit values* are referred to as *spins*. Moreover, instead of taking the values 0 or 1, spins are considered to take the values -1 or +1. The analogy is with magnetic spin systems in physics, in which we have a number of interacting particles, each with a different spin. Consequently, the sum of the spin values in a string lies in the range $-\ell \leq x \leq \ell$. This sum (which in our examples is the number of ones, or *unitation*) is called the *magnetization* of the string. In physics, the concern is with how such systems relax into some low energy state. The GA is therefore usually discussed from a fitness *minimization* viewpoint. For example, the basin with a barrier function is usually drawn upside down, and is sometimes called 'two wells'. The notion of the *entropy* of a fitness (or magnetization) value is also used. This refers to the fact that some magnetization values correspond to a small set of strings (low entropy) and others to a large set of strings (high entropy). For example, there is only one string with magnetization ℓ, but there are many strings with zero magnetization (assuming bits take on the values ±1). This may arise in a discussion of the effect of mutation, say, in which it is said that mutation tends to drive the population towards high entropy regions of the search space.

7.7. BIBLIOGRAPHIC NOTES

Exercises

1. Given a probability generating function
$$G(x) = \sum_{k \geq 0} \Pr[X = k] x^k$$
we know that $E[X] = G'(1)$. Show that the variance is
$$V[X] = G''(1) + G'(1) - [G'(1)]^2$$

2. Consider the fixed point equation for *Onemax* derived at the end of section 7.6. Show that if mutation is set to $u = 1/\ell$ then the average fitness at the fixed point is approximately two-thirds of the string length.

3. Consider the needle-in-a-haystack function on bit strings of length ℓ:
$$f(x) = 1 + [\boldsymbol{x} = \boldsymbol{0}]a$$
where $a > 0$ gives the height of the needle. Let $\pi(t)$ be the proportion of the population on the needle (i.e. the string of all zeros) at time t.

   (a) Show that the effect of proportional selection is
$$\pi(t+1) = \frac{(1+a)\pi(t)}{1 + a\pi(t)}$$

   (b) Given a bitwise mutation rate of u, what is the probability that a string of all zeros is not destroyed?

   (c) Approximate the effect of mutation by assuming that the probability of creating the string of all zeros is negligible. Write down the evolution equation combining the effects of selection and mutation.

   (d) If mutation is assumed to be small, show that the fixed point of the evolution equation is approximately
$$1 - u\ell\left(1 + \frac{1}{a}\right)$$

4. (continuation) Analyse the needle-in-a-haystack problem for binary tournament selection. Assuming that mutation is very small and that the chance of creating copies of the needle are negligible, show that at the fixed point the proportion of the population on the needle is approximately
$$2 - (1-u)^{-\ell}$$

5. Analyse the effect of binary tournament selection for *Onemax*, by assuming that the fitness distribution is Gaussian.

Chapter 8

Predicting GA Performance

From relatively early in the development of theoretical ideas about genetic algorithms, an important aspect has been the delineation of classes of problems for which GAs were thought to be especially well (or ill) suited. We see this first in Goldberg's concept of deception [97]. Partly this was driven by the need to verify the building block hypothesis, at least in a negative way, as we have seen in Chapter 3. It was about this time that the notion of problems being 'GA-hard' or 'GA-easy' began to be used. Although the terminology recalls that of computational complexity, there were no precise definitions of what was meant by such terms.

Later, deception—as originally conceived—was shown to be neither necessary or sufficient [107] to characterise a problem as GA-hard. Other types of problem were investigated, some involving massive multi-modality or long paths to a single global optimum. However, the multiple local optima and the long paths usually related either to continuous space, or to the Hamming landscape of a discrete optimization problem, while its relevance to how a GA solved such a problem was not entirely clear. (More will be said about this in Chapter 9.) Nevertheless, there is now a considerable literature on the question of characterizing and predicting GA performance.

8.1 Introduction

In considering this question, there are two related aspects that place the emphasis slightly differently. Firstly, we can ask questions about the general tendency of GA behaviour, with the idea of being able to identify classes of problems that GAs find either 'easy' or 'hard'. It was these questions that were addressed in studies on deception and building blocks. However, even should such a categorization be possible, it still leaves open the second question of how we can recognize *a priori* that the particular *instance* that we wish to solve belongs to one class or the other. In fact, there is a sense in which the second question is more important than the first, in that if we could characterize and recognize the difficulty of problem instances in some way, we could use the methodology in defining classes of hard or easy problems.

8.2 Epistasis Variance

The first serious attempt to answer the second of these questions was by Davidor [45, 46]. The central concept in these and subsequent studies is the idea of *epistasis*. This idea has already been defined in general terms in Chapter 1—the assumption that fitness is not related to genes in an additive way. It can be interpreted as expressing non-linearity in the fitness function, and in seeking to measure it, the assumption is that, at least approximately, the more epistatic the problem is, the harder it is to find an optimum.

Exactly what fitness means for real organisms is—as noted in Chapter 1—a complex problem, but for the case of GAs it is conceptually a simple matter. The epistatic nature of many familiar functions (such as those described in Appendix A) can be determined by inspection. For unknown functions it is harder, but were we to know the fitness of every phenotype in the Universe of all possible phenotypes, we could in principle calculate the effect on fitness of any combination of genes. Of course, if we really did know the Universe, regarding it as an *optimization* problem would be pointless anyway, but there is at least a theoretical possibility that we might be able to characterize a function using only a sample of points.

If this could be achieved, and the *degree* of epistasis in a given problem determined, we might be able to decide on the most suitable strategy for solving it. At one end of the spectrum, a problem with very little epistasis should be easily solved by a suitable linear or quasi-linear numerical method. At the other end, a highly epistatic problem is unlikely to be solvable by any systematic method, including a GA. Problems with intermediate epistasis might be those worth attempting with a GA. This, at least, appears to be the assumption that has prompted the investigations that we shall consider in this chapter.

Another aspect of the investigation of epistasis is also intrinsically relevant to the study of GAs. As we have seen, GAs tend to place greater emphasis than other techniques on the question of representation, and a particular choice of coding may render a simple linear function epistatic. Conversely, by choosing a different encoding function, it may be possible to reduce the degree of epistasis in a problem. It would clearly be valuable to be able to measure and compare the epistasis existing in different codings of the same problem. Related to this is the question of the choice of operators: as we shall see in Chapter 9, it is not only the fitness function and the problem encoding that influence its difficulty. However, we shall postpone questions of this nature for the next chapter.

8.3 Davidor's Methodology

Davidor assumed populations of binary strings $\{x\}$ of length ℓ, for which he defines several quantities. The basic idea is that for a given population $P(t)$ of size N, the average fitness value can be determined as

$$\bar{f} = \sum_{x \in P(t)} f(x)/N$$

Subtracting this value from the fitness of a given string x produces the excess string fitness value

$$E(x) = f(x) - \bar{f}$$

This simply re-states the fitness of all strings in $P(t)$ with respect to the population average.

Now some allele values will occur more frequently than others at each locus. We may count the number of occurrences of allele a for each gene i, denoted by $N_i(a)$, and compute the average allele value

$$A_i(a) = \sum_{x:x_i=a} f(x)/N_i(a)$$

The excess allele value measures the effect of having allele a at gene i, and is given by

$$E_i(a) = A_i(a) - \bar{f}$$

The excess genic value of string x is the value obtained by summing the excess allele values at each gene,

$$E_G(x) = \sum_{i=1}^{\ell} E_i(a)$$

Finally, the epistasis value is the difference between the actual value of string x and the genic value predicted by the above analysis, which reduces to

$$\epsilon(x) = E(x) - E_G(x)$$

Thus, the value $\epsilon(x)$ measure the discrepancy between the actual fitness and a 'linear' model of fitness.[1] Davidor went on to define a measure of 'epistasis variance' that unfortunately proved somewhat difficult to interpret. In fact, his methodology has a basic flaw, as we shall see later.

8.4 An Experimental Design Approach

Reeves and Wright [217, 218] pointed out that what Davidor was attempting was in fact exactly what statisticians have been doing in the sub-discipline known as *experimental design* (ED).[2]

Davidor was implicitly assuming a linear model (defined on the genes) for the fitness of a string. We can express this model as

$$f(x) = \text{constant} + \sum_{i=1}^{\ell}(\text{effect of allele at gene } i)$$

$$+ \sum_{i=1}^{\ell-1} \sum_{j=i+1}^{\ell} (\text{interaction between alleles at gene } i \text{ and gene } j)$$

$$+ \ldots$$

$$+ (\text{interaction between alleles at gene 1, gene 2, \ldots, gene } \ell)$$

$$+ \text{random error}$$

In conventional experimental design, the above model would actually be written in parametric form. For example, the model for a string of 3 genes could be written as follows:

$$f_{pqrs} = \mu + \alpha_p + \beta_q + (\alpha\beta)_{pq} + \gamma_r + (\alpha\gamma)_{pr} + (\beta\gamma)_{qr} + (\alpha\beta\gamma)_{pqr} + \epsilon_{pqrs} \quad (8.1)$$

where f_{pqrs} is the fitness of the string (p, q, r), and the subscript s denotes the replication number (i.e. the sth occurrence of the string), and allows for stochastic fitness evaluation. If there are no stochastic effects, we can of course drop the subscript s. The parameters on the right-hand side are as follows:

---

[1] Note that we have merely followed Davidor's original notation here; these $E()$ values don't necessarily have anything to do with expectations.

[2] It is interesting that it was Sir Ronald Fisher who played a major part in the development of this area of statistics, just as he did in the case of population genetics, and both have a place in the analysis of GAs.

8.4. AN EXPERIMENTAL DESIGN APPROACH

μ	average fitness
α_p	effect of allele p at gene 1
β_q	effect of allele q at gene 2
$(\alpha\beta)_{pq}$	interaction of allele p at gene 1 and allele q at gene 2
γ_r	effect of allele r at gene 3
$(\alpha\gamma)_{pr}$	interaction of allele p at gene 1 and allele r at gene 3
$(\beta\gamma)_{qr}$	interaction of allele q at gene 2 and allele r at gene 3
$(\alpha\beta\gamma)_{pqr}$	interaction of alleles p at gene 1, q at gene 2 and r at gene 3
ε_{pqrs}	random error for replication s of string (p, q, r)

In many, although not all, applications of GAs random error is not considered relevant, and we merely note that this formulation allows it to be treated if desired, although in what follows we shall assume the problems to be deterministic.

8.4.1 An example

Suppose we have a string with $\ell = 3$, and that the fitness of every string in the Universe is known. There are of course $2^3 = 8$ strings, and therefore 8 fitness values, but the experimental design model above has 27 parameters. It is thus essential to impose some side conditions if these parameters are to be estimated; the usual ones are the obvious idea that the parameters represent deviations from a central value. This means that at every order of interaction, the parameters must sum to zero for each subscript. This results in an additional 19 independent relationships such as

$$\sum_p \alpha_p = 0$$
$$\sum_p (\alpha\beta)_{pq} = 0 \quad \text{for } q = 0, 1$$
$$\sum_p (\alpha\beta\gamma) = 0 \quad \text{for } q, r = 0, 1$$

and so on. Imposing these side conditions allows the 'solution' of the above model—in the sense that all the parameter values can be determined if we have observed every one of the 8 possible strings. For example, we find that

$$\mu = f_{***}$$
$$\mu + \alpha_p = f_{p**} \quad \text{for } p = 0, 1$$
$$\mu + \beta_q = f_{*q*} \quad \text{for } q = 0, 1$$
$$\mu + \gamma_r = f_{**r} \quad \text{for } r = 0, 1$$

Table 8.1: Analysis of Variance Table

Source of variation	Degrees of freedom	Sum of squares (SS)	Davidor's notation
Between alleles of gene 1	1	$\sum_{pqr}(f_{p**} - f_{***})^2$	$\sum[E_1(a)]^2$
Between alleles of gene 2	1	$\sum_{pqr}(f_{*q*} - f_{***})^2$	$\sum[E_2(a)]^2$
Between alleles of gene 3	1	$\sum_{pqr}(f_{**r} - f_{***})^2$	$\sum[E_3(a)]^2$
Total main effects (i.e.,'genic' effect)	3	sum of above	sum of above
Interactions (i.e.,epistatic effect)	4	$\sum_{pqr}(f_{pqr} - f_{p**} - f_{*q*} - f_{**r} + 2f_{***})^2$	$\sum[\epsilon(x)]^2$
Total	7	$\sum(f_{pqr} - f_{***})^2$	$\sum[E(x)]^2$

where the notation f_{p**}, for instance, means averaging over subscripts q and r. The effects can be seen to be exactly equivalent to Davidor's 'excess allele values' as defined above. For instance, his $A_1(p) = f_{p**}$, so that $E_1(p) = \alpha_p$. Similarly, his 'excess genic values' $E(A)$ are found by summing α_p, β_q and γ_r for each possible combination of p, q, r. Finally, his 'string genic value' is clearly

$$\mu + \alpha_p + \beta_q + \gamma_r.$$

The difference between the actual value and the genic value, $\epsilon(S)$, is therefore simply the sum of all the interaction terms; in other words, zero epistasis is seen to be equivalent to having no interactions in the model.

8.4.2 Analysis of Variance

The normal procedure in experimental design is to perform an 'Analysis of Variance' (Anova), whereby the variability of the fitness values (measured by sums of squared deviations from mean fitness, and denoted by SS) is partitioned into orthogonal components from identifiable sources. In Table 8.1 we give a conventional Anova table for our 3-bit example, with Davidor's notation alongside.

8.5. WALSH REPRESENTATION

The *degrees of freedom* in this table are the number of independent elements in the associated SS; for example, in the Total SS term, only 7 of the $(f_{pqr} - f_{***})$ terms are independent, since they must satisfy the relationship

$$\sum_{pqr}(f_{pqr} - f_{***}) = 0$$

It is easy to prove (as we shall see in Section 8.5.1) that

Total SS = Main effects SS + Interaction SS.

Since Davidor has simply divided these values by a constant to obtain his 'variances', it is hardly surprising that he finds that

Total 'variance' = 'Genic variance' + 'Epistasis variance'

8.5 Walsh representation

As shown in [217], for the case of binary strings, this way of decomposing the fitness of all the strings in a Universe is equivalent to the Walsh function decomposition described in Chapter 3. For 3-bit strings, we find[3]

$$\mu = w_0$$
$$\alpha_0 = w_1$$
$$\beta_0 = w_2$$
$$(\alpha\beta)_{00} = w_3$$
$$\gamma_0 = w_4$$
$$(\alpha\gamma)_{00} = w_5$$
$$(\beta\gamma)_{00} = w_6$$
$$(\alpha\beta\gamma)_{000} = w_7$$

The 'mapping' from the Walsh coefficient indices to the appropriate 'effect' is given by writing the effects in what is known in experimental design as *standard order*: in this case

$$\{\mu, \alpha, \beta, \alpha\beta, \gamma, \alpha\gamma, \beta\gamma, \alpha\beta\gamma\}.$$

---

[3] In experimental design it is normal to list the effects from left to right, whereas in most of the literature on Walsh functions, the convention is to read the bits from right to left. Of course, the labelling is arbitrary, and the necessary adjustment is quite straightforward.

The general pattern is fairly obvious—on adding another factor the next set of effects is obtained by 'combining' the new factor with the effects already listed, in the same order. Thus in the case of a 4-bit problem, for example, the next 8 in order will be

$$\{\delta, \alpha\delta, \beta\delta, \alpha\beta\delta, \gamma\delta, \alpha\gamma\delta, \beta\gamma\delta, \alpha\beta\gamma\delta\}.$$

Another way of seeing this equivalence is as follows: the 'interaction level' of a particular Walsh coefficient is given by the number of 1s in the binary equivalent of its subscript, and the specific factors involved are indicated by the positions of those 1s. For example, consider the coefficient w_{13} in a 4-bit problem: the binary equivalent of 13 is 1 1 0 1, so this represents a 3-factor interaction, and the factors participating in this interaction are numbers 1, 3 and 4. Therefore w_{13} measures the effect of the interaction $(\alpha\gamma\delta)$ (using the right-to-left labelling convention). The sign of the coefficient is of course determined by the relevant value of x as described in Chapter 3. For example, the sign of the $(\alpha\gamma\delta)$ component in the value of f for $x = (0\,1\,0\,1)$ would be found from

$$\varpi(0\,1\,0\,1 \wedge 1\,1\,0\,1) = \varpi(2) = 1$$

(where ϖ is the parity function), while that of the corresponding component for $x = (0\,0\,0\,1)$ would be

$$\varpi(0\,0\,0\,1 \wedge 1\,1\,0\,1) = \varpi(1) = -1$$

In what follows, we shall use whichever notation is appropriate for the point we are making. While Walsh notation has advantages in terms of compactness and general mathematical convenience, the interpretation of the coefficients as ED effects is perhaps more transparent. The validity of this point was demonstrated in [217], where an analysis of Goldberg's conditions for deception from an ED perspective revealed an error in his original specification. (Two inequalities in [98] which *should* have read $w_3 + w_5 > w_1 + w_7$ and $w_3 + w_6 > w_2 + w_7$, had had their right-hand sides transposed.)

8.5.1 Davidor's 'variance' measure

We can partition the Walsh coefficients into subsets defined by their order of interaction.

$$\begin{aligned} I_0 &= \{w_0\} \\ I_1 &= \{w_1, w_2, w_4, \ldots\} \\ I_2 &= \{w_3, w_5, w_6, \ldots\} \\ I_3 &= \{w_7, w_{11}, w_{13}, \ldots\} \end{aligned}$$

8.5. WALSH REPRESENTATION

and so on, with

$$I = \bigcup_{r=0}^{\ell} I_r$$

Thus the linear approximation of $f(\boldsymbol{x})$ is

$$\xi(\boldsymbol{x}) = w_0 + \sum_{j \in I_1} w_j \psi_j(\boldsymbol{x}) \tag{8.2}$$

Davidor's 'variance' measure, as modified by Reeves and Wright [218], can then be written (suppressing subscripts and arguments for clarity) as follows:

Definition 8.1 *The epistasis variance η of a function f is the fraction of the total variation in the values that f can attain that is unexplained by its linear approximation:*

$$\eta = \frac{\sum (f - \xi)^2}{\sum (f - \bar{f})^2} \tag{8.3}$$

where $\bar{f}$ is the mean value of all strings in the Universe and the sum is over all strings.

In fact, Davidor used only the numerator of this quantity; however, it is clearly in need of some form of normalization. One possibility [282] is first to standardize all f values by dividing by their maximum value. Another solution [180] is to divide by $\sum f^2$. However, while these would guarantee that its value lies in the range $[0, 1]$, the interpretation of such a measure is not clear. Furthermore, the value would not be invariant to a translation (i.e., a change in origin), which is clearly undesirable. From an ED perspective there is only one sensible normalization—to divide by the overall sum of squared deviations from the mean. The value η as given above is then translation-invariant, and has the simple interpretation that it measures the fraction of the total 'variance' not explained by linear effects. Thus, we might expect that a value near to 0 expresses the fact that the function is close to linear, while a value near to 1 indicates a function that is highly non-linear. For example, for a 3-bit problem, we have

$$\eta = \frac{\sum_{pqr}[f_{pqr} - (\mu + \alpha_p + \beta_q + \gamma_r)]^2}{\sum_{pqr}(f_{pqr} - f_{***})^2}$$

and we see that the numerator reduces to

$$\sum_{pqr}[(\alpha\beta)_{pq} + (\alpha\gamma)_{pr} + (\beta\gamma)_{qr} + (\alpha\beta\gamma)_{pqr}]^2.$$

In fact, this can be further reduced and indeed generalized. Standard ED theory (for example, [119] or [172]) shows that sums of squared deviations can be additively partitioned, as we mentioned above. We can demonstrate the result in this context as follows:

Proposition 8.1 *Davidor's variance measure can be written as the sum of the squared interaction effects; i.e.*

$$\eta = \frac{\sum (\text{interaction effect})^2}{\sum (f - \bar{f})^2} \quad (8.4)$$

where the sum is over all strings.

Proof From Equation (8.3) the numerator can be written as

$$\sum_{pqr\ldots} [(\alpha\beta)_{pq} + (\alpha\gamma)_{pr} + (\beta\gamma)_{qr} + (\alpha\beta\gamma)_{pqr} + \cdots]^2.$$

We can expand about the first component of the sum to obtain

$$\sum_{pqr\ldots} (\alpha\beta)^2_{pq} + [(\alpha\gamma)_{pr} + (\beta\gamma)_{qr} + (\alpha\beta\gamma)_{pqr} + \cdots]^2$$
$$+ 2(\alpha\beta)_{pq}[(\alpha\gamma)_{pr} + (\beta\gamma)_{qr} + (\alpha\beta\gamma)_{pqr} + \cdots]$$

Now every term in the cross-product (the second line in the above formula) will have a subscript that can be summed over, splitting the cross-product into two parts, one of which is zero because of the side conditions. For instance (suppressing the other subscripts for the sake of clarity),

$$\sum_{pqr}(\alpha\beta)_{pq}(\alpha\gamma)_{pr} = \sum_p\sum_q\sum_r (\alpha\beta)_{pq}(\alpha\gamma)_{pr} = \sum_p\sum_q(\alpha\beta)_{pq}\sum_r(\alpha\gamma)_{pr} = 0$$

or again,

$$\sum_{pqr}(\alpha\beta)_{pq}(\beta\gamma)_{qr} = \sum_q\sum_r(\beta\gamma)_{qr}\sum_p(\alpha\beta)_{pq} = 0.$$

Thus the cross-product terms disappear; a procedure that can be repeated until all the terms in the numerator are simply squared interaction effects. □

Note (1) In the Walsh representation, the epistasis variance measure can be written as

$$\eta = \frac{\sum_{j \in I \setminus \{I_0 \cup I_1\}} w_j^2}{\sum_{j \in I \setminus I_0} w_j^2} \quad (8.5)$$

8.5. WALSH REPRESENTATION

Note (2) An immediate corollary of this result is as stated in [218]: to every function f with a given value of η, there will be many functions whose decomposition differs from f only in the sign of the interaction effects. Clearly such functions cannot be distinguished by the epistasis variance measure.

Note (3) The idea inherent in the definition of η can be extended to epistasis of a particular order. We can define the kth order approximation of f as

$$\xi_k(\boldsymbol{x}) = w_0 + \sum_{j \in J_k} w_j \psi_j(\boldsymbol{x})$$

where

$$J_k = \bigcup_{r=1}^{k} I_r,$$

and the corresponding kth order epistasis variance as

$$\eta_k = \frac{\sum (f - \xi_k)^2}{\sum (f - \bar{f})^2},$$

where $\eta = \eta_1 \geq \eta_2 \geq \ldots \geq \eta_\ell = 0$. In other words, the sequence of η values measures the effect of successively better approximations to the function f. It is precisely this idea that underlies the Analysis of Variance described in Section 8.1. We could further consider the *differences* $(\eta_j - \eta_{j+1})$ as measuring the *components of variance*: the amount of variance due to interactions of order $j+1$. (For convenience, we may define $\eta_0 = 1$ in order to extend this formula to all values of j.)

It might seem therefore that this idea provides an excellent way of distinguishing between functions with different levels of epistasis, but there are some problems that cannot easily be resolved.

8.5.2 Signs matter

Assuming for the moment that we can obtain a good estimate of the epistasis variance, its value still does not contain enough information to say whether the problem is hard or easy. In [218] the importance of the *signs* of the interaction effects is pointed out. If the interaction acts counter to the joint influence of its associated main effects, we have essentially what we have previously called 'deception'. On the other hand, no matter how large an interaction effect is, if its sign reinforces the message of its associated main effects, there is no deception. Figure 8.1 illustrates this point.

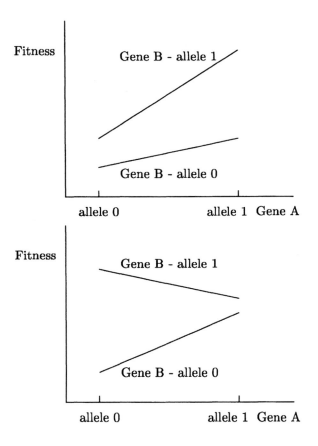

Figure 8.1: Benign and malign interactions. Note that the lines are not meant to imply continuity here—they are merely to emphasize the direction of the interactions.

8.5. WALSH REPRESENTATION

In the upper diagram of this figure, each gene should clearly be set to have allele 1. Epistasis exists, since the joint effect of having A and B set at 1 exceeds the sum of the individual main effects, but the interaction merely reinforces the main effects. In the lower diagram, the interaction has a malign influence: the best allele for genes A and B considered individually is 1, but overall it is better to set gene A at 0 and B at 1. This would be shown by an interaction that is negative for the combination (1,1). However, the value η uses only the square of the interaction effects, so it cannot distinguish between cases where epistasis helps and where it harms. A high value of η would certainly indicate small main effects, but if the interaction effects had the 'right' signs, it could still be a trivial problem to solve.

8.5.3 Implications of sampling

As if this was not in itself a fatal blow to hopes of using η as a practical tool for measuring epistasis, there is a further problem. The whole analysis so far has been predicated on knowledge of the complete Universe. In reality, if we knew the Universe, there would be no point in using an optimization algorithm, so the value of epistasis variance would be of merely academic importance. Naturally we don't really know the Universe, as we are typically dealing with an enormous search space, so the temptation is clearly simply to take a random sample and hope that η is still in some sense meaningful.

Unfortunately, the meaning of η depends not only on the size of the sample, but also on its composition. Reeves and Wright [218] argued that in principle it is not possible to estimate the value of the effects independently of each other from a sample. All that we can do is to estimate some linear combination of the effects, a combination that depends upon the composition of the sample used.

Table 8.2 illustrates the problem in the case where half the Universe is known. For example, from Equation (8.2) the true value of α_1 would be

$$\frac{f_1 + f_3 + f_5 + f_7 - f_0 - f_2 - f_4 - f_6}{8}$$

while the obvious *estimates* of both α_1 and β_1 from the half-fraction (we denote these estimates by the $\hat{}$ symbol) are

$$\frac{f_3 + f_7 - f_0 - f_4}{4}$$

These clearly cannot be distinguished—statisticians talk of these effects being 'aliased' or 'confounded'. Comparing this with the true values, it is easy

Table 8.2: A half-fraction of the Universe. The estimates $\hat{\alpha}_1$ and $\hat{\beta}_1$, for example, are indistinguishable.

fitness value	String	
	Universe	Half-fraction
f_0	0 0 0	0 0 0
f_1	0 0 1	
f_2	0 1 0	
f_3	0 1 1	0 1 1
f_4	1 0 0	1 0 0
f_5	1 0 1	
f_6	1 1 0	
f_7	1 1 1	1 1 1

to see that
$$\hat{\alpha}_1 = \alpha_1 + \beta_1$$

In other words, all we can really do is estimate linear combinations of some effects. What these combinations might be will be determined by the particular sample we are using. In practice, we may hope that most of the interaction effects (especially the higher-order ones) are negligible, so that the sample estimates we obtain will reflect the true values fairly well. Actually, this assumption is at the root of most applications in experimental design. However, in ED the sample would not be a random one, rather it would be constructed on principles of orthogonality so that the task of estimating the values of the effects that are assumed to be non-negligible is easier. (That is, we would take care to confound the effects we are interested in only with those we assume to be negligible. We will say a little more about this in the discussion later.) In the case of experimental design, the Universe is usually rather smaller than is likely in GA applications, and in the latter case, constructing an orthogonal subset is therefore a rather more difficult task than it generally is for ED. The temptation therefore is just to rely on large random samples, but the sample size may need to be large indeed if the baleful effects of confounding are to be eradicated.

8.5.4 Still NP-hard

Merely to obtain a reliable indicator of a function's degree of epistasis is not usually an end in itself. The purpose (whether expressed or implied) is to use this information to help us find the optimum. Whether we would still wish to use a GA is in itself doubtful. Having obtained (at some computational cost) a substantial amount of knowledge of the function, it would seem bizarre to throw it away and start again with an algorithm that ignores this information. The 'obvious' thing to do would be to try to use this information directly to determine the optimal solution. But how can we use such information?

There is yet another difficulty here. Suppose we eventually achieve reliable estimates of all the non-negligible effects; if we assume maximization as the goal, we would like to be able to give + signs to all the positive main effects, and − signs to the negative ones. This is trivial if there is no epistasis. Unfortunately, if the function is epistatic, whether it works will depend on how malign the epistasis is. Giving a + sign to two main effects, for example, might entail a negative influence for their associated interaction, so the allocation of signs is not straightforward. Essentially we have a version of the *satisfiability* problem to solve—an assignment of Boolean values has to be found that will maximize some weighted sum of 'clauses'— and of course such problems are NP-hard [88]. As explained by Heckendorn et al. [118], unless P = NP, even an exact Walsh analysis for a problem of bounded epistasis (i.e. an upper limit on the number of factors that can interact) is insufficient to guarantee inferring the global optimum in polynomially bounded time. Naudts [178] has also pointed this out, showing that this is the case even if the interactions are restricted to be no more than second-order.

8.6 Other related measures

Rochet [238] suggested the use of a quantity that she called *epistasis correlation*, defined as

$$\varrho = \frac{\sum (f - \bar{f})(\xi - \bar{\xi})}{\sqrt{\sum (f - \bar{f})^2 \sum (\xi - \bar{\xi})^2}} \tag{8.6}$$

where ξ is the linear approximation to f given by Equation (8.2) (subscripts have again been suppressed for clarity). Of course, in practice this entails first making estimates of the relevant Walsh coefficients in order to obtain ξ. A family of values $\{\varrho_j\}$ for different orders of approximation can be defined analogously, just as they were for η. In fact, ϱ is closely related to η, as in the following (similar results can be proved for ϱ_j):

Proposition 8.2

$$\varrho = \sqrt{1-\eta}$$

Proof Note, firstly, that Equation (8.3) can be formulated as

$$\eta = \frac{\sum(f-\xi)^2}{\sum(f-\bar{f})^2},$$

and secondly, that the effect of the side conditions on the effects is that

$$\bar{\xi} = \bar{f} = \mu.$$

Thus

$$f - \xi \equiv (f - \bar{f}) - (\xi - \bar{\xi}).$$

Equation (8.3) thus becomes

$$\begin{aligned}\eta &= \frac{\sum(f-\bar{f})^2 + \sum(\xi-\bar{\xi})^2 - 2\sum(f-\bar{f})(\xi-\bar{\xi})}{\sum(f-\bar{f})^2} \\ &= 1 + \frac{\sum(\xi-\bar{\xi})^2}{\sum(f-\bar{f})^2} - 2\varrho\sqrt{\frac{\sum(\xi-\bar{\xi})^2}{\sum(f-\bar{f})^2}}.\end{aligned}$$

However, because of the additive partitioning of the sums of squared deviations

$$\sum(f-\bar{f})^2 = \sum(f-\xi)^2 + \sum(\xi-\bar{\xi})^2$$

and it follows that

$$\frac{\sum(\xi-\bar{\xi})^2}{\sum(f-\bar{f})^2} = 1 - \eta$$

which is of course just another way of writing Equation (8.3). Simple algebra now shows that

$$2\varrho\sqrt{1-\eta} = 2(1-\eta)$$

and the result follows. □

The consequences of this result are clear. Firstly, ϱ gives us nothing that is not already available from η. Secondly, it suffers from the same basic problem as η—it is equally incapable of distinguishing between cases of benign and malign epistasis, so its value need not be indicative of problem difficulty. The situation with respect to sampling is also as bad. Indeed, it could be argued that the need to *estimate* $\mu, \alpha_p, \ldots$ before ϱ can be calculated

8.6. OTHER RELATED MEASURES

actually makes the situation worse. Certainly, if there is high epistasis in the problem, a large sample size will be needed or else there is a high risk of confounding the estimates of the main effects with some combination of interactions, which would mean the 'linear model' ξ is biased, and the epistasis correlation obtained is seriously compromised.

Heckendorn et al. [116, 117] have proposed using Walsh sums as an alternative way of specifying the amount of epistasis. These are based on the absolute values of the Walsh coefficients relating to a given order of interaction:

$$\varsigma_k = \sum_{j \in I_k} |w_j|.$$

The relative size of these coefficients can be used in an analogous way to the sums of squares for η, although without their nice decomposition properties. However, it is clear that Walsh sums face exactly the same problem as the epistasis variance approach—they are the same modulo a change in sign of the Walsh coefficients, and thus cannot guarantee to tell us anything about problem difficulty. (It is fair to point out that *a posteriori* identification of problem difficulty was not the main focus of the work in [116, 117].)

8.6.1 Static ϕ

Heckendorn et al. [116] also proposed a measure related to schema theory, which they called 'static ϕ'. The purpose of this measure was to assess the consistency with which schema fitness information matches string fitnesses. For instance, if the string (1 1 1) is optimal, then the order-2 schema fitness averages are consistent with this fact if

$$f_{11*} > \left\{ \begin{array}{c} f_{10*} \\ f_{01*} \end{array} \right\} > f_{00*}$$

The static ϕ measure relative to a string S and order-k schemata $H_1, \ldots, H_{2^k}$ is given by

$$\sum_{i,j} [f_{H_i} > f_{H_j}] \max\left(m(H_i, S) - m(H_j, S), 0\right)$$

where $m(H_i, S)$ is a 'match' function, which counts the number of loci where S and H_i agree (e.g., $m(1\ 1\ *, 1\ 1\ 1) = 2$), and $[expr]$ is the indicator function notation introduced in Chapter 2. A combined measure of difficulty can be obtained by summing over all partitions (sets of schemata of the same order). Rana [208] shows that this measure can be quite good at predicting the ease or difficulty of finding the global optimum. This is fine if we know where the global optimum is—but if we did know it, perhaps we would not be very interested in the abstract question of how difficult it was to find it. Also, of course, if we wish to measure consistency across all partitions, we need schema averages of all orders, which (again!) means we need to know the Universe before we can properly compute ϕ.

8.6.2 Fitness-distance correlation

An obvious criticism of the epistasis variance approach is that it completely ignores the question of the search algorithm, and simply looks at the fitness function. The same is true of the approach of Rosé et al. [239], who focus on the 'density of states'—the probability distribution of fitness *values*. Such information may well be useful, but it says nothing about how those fitness values are topologically related to each other: do they form a smooth continuum[4] building towards a solitary peak in the 'landscape' (think of Mount Fuji), or do they form a spiky pattern of many isolated peaks (think of the Dolomites)? In fact, this will depend not only on the fitness function, but also on the search algorithm, as we shall see at length in Chapter 9.

In the light of this, Jones and Forrest [137] suggested measuring the correlation between fitness and distance (ρ_{fd}) of sets of points on a landscape induced by the application of a specified operator. This is given by

$$\rho_{fd} = \frac{\sum (f - \bar{f})(d - \bar{d})}{\sqrt{\sum (f - \bar{f})^2 \sum (d - \bar{d})^2}} \qquad (8.7)$$

where d is the distance of each string from its nearest global maximum, and again the sum is over all strings in the Universe. With ρ_{fd}, some concept of distance between solutions is now brought into play, and we can more properly use the term *landscape*. However, which distance is appropriate? The most commonly used distance measure to use for binary strings is the Hamming distance, as defined in Chapter 2. However, this only defines one landscape; other distance measures will define different landscapes, and it is certainly not obvious that the Hamming landscape is the most appropriate one for the analysis of GAs, for example. As is pointed out in [6, 122], mutation is in some sense an approximation to the Hamming landscape, in that the effect of mutation at typically applied rates is that although offspring-parent distance is technically a random variable, in practice it is

---

[4]Of course, smoothness and continuity are not really properties of discrete spaces, but as long as we understand this, such metaphors can be helpful.

8.6. OTHER RELATED MEASURES

often $d_H(x, y) = 1$ or (occasionally) 2; only very rarely is it greater than 2. However, the offspring that can be generated by crossover may be a long way from their parents in terms of Hamming distance.

In any case, ρ_{fd} also suffers from some severe problems. In the first place, the global optimum is usually unknown, so the values of d are impossible to find. In practice, we could use the fittest known string as a surrogate for the global optimum, but there is no guarantee that the value of ρ_{fd} thus obtained will tell us anything very useful, unless the surrogate is close to the true optimum. On the positive side, examples of this approach in operational research [25, 163, 224] have shown that correlations obtained in this way are often significant, even when the distance measure is not exactly based on the operators used. There is thus a pragmatic argument for building algorithms that can exploit the underlying landscape structure that seems to emerge 'naturally' for many COPs. But it is still not clear that GAs necessarily induce such structure, so that measuring ρ_{fd} may not tell us much in general about the difficulty of a problem for a recombinative GA, in spite of the specific evidence in [137]. Indeed, several counter-examples have been reported [6, 197, 265].

Secondly, as with η, very different functions can give the same ρ_{fd} value. In order to demonstrate this fact, suppose that we create a new function using a quantity Δ and a partition of the Universe of strings into two subsets, Z^+ and Z^-, where

$$f(z) \mapsto f(z) + \Delta; \; z \in Z^+$$
$$f(z) \mapsto f(z) - \Delta; \; z \in Z^-.$$

These strings subdivide further as follows:

$$Z^+ \text{ (resp. } Z^-) = \bigcup_{d=0}^{\ell} Z_d^+ \text{ (resp. } Z_d^-)$$

where d is the distance from z to the global maximum. We use the notation n_d^+ (resp. n_d^-) $= |Z_d^+|$ (resp. $|Z_d^-|$). (Clearly $n_d^+ + n_d^- = \binom{n}{d}$ $\forall d$.)

Proposition 8.3 *Given a function $f : \{0,1\}^\ell \mapsto \mathbb{R}$ with a single global maximum at the string z^*, and the transformation of f as described above, the value of ρ_{fd} is unaffected by such a transformation provided that it satisfies the following conditions:*

1. $$\Delta = \frac{2}{n}\left(\sum_{z \in Z^-} f(z) - \sum_{z \in Z^+} f(z)\right) > 0$$

Table 8.3: Two 3-bit problems with the same ρ_{fd} value. The maximum is assumed to be at (1 1 1).

Original	String	Transformed	d
f_0	0 0 0	$f_0 - \Delta$	3
f_1	0 0 1	$f_1 + \Delta$	2
f_2	0 1 0	$f_2 + \Delta$	2
f_3	0 1 1	$f_3 - \Delta$	1
f_4	1 0 0	$f_4 + \Delta$	2
f_5	1 0 1	$f_5 - \Delta$	1
f_6	1 1 0	$f_6 - \Delta$	1
f_7	1 1 1	$f_7 + \Delta$	0

2. $\quad z^* \in Z^+$ or, equivalently, $n_0^+ = 1$

3. $\quad \sum_{d=0}^{\ell} d(n_d^+ - n_d^-) = 0$

4. $\quad \sum_{d=0}^{\ell} (n_d^+ - n_d^-) = 0$

Proof In the first place, because we choose to *increase* its fitness by Δ, z^* is still the optimal string (condition 2). It can then easily be verified that the value of $\sum (d - \bar{d})^2$ is unaffected by the transformation. The fact that z^* is still the maximum, together with condition 3 and the fact that $\bar{f}$ is also unchanged (condition 4) then ensures that the numerator of Equation (8.7) remains the same. Finally, condition 1 will ensure that the denominator is also unchanged by the transformation. Thus the value of ρ_{fd} is the same for the new function. $\square$

An example of such a transformation is given in Table 8.3.

Note (1) Whether there is always a solution that satisfies these conditions for all ℓ is unknown; however, a solution has always been found for cases examined so far. These conditions are also stricter than they need to be: even if $\Delta \leq 0$, the status of z^* as the maximum may not change, and the other conditions will still be sufficient.

Note (2) Further relaxations are also possible. For example, it is only a little more difficult to find conditions for two values Δ_1 and Δ_2 and make similar stipulations regarding them. In fact, since one of Δ_1 and Δ_2 can then be chosen freely, we could find arbitrarily many functions having the same value of ρ_{fd}. Again, whether the necessary conditions can be satisfied for all ℓ is not known.

Note (3) This type of change is precisely that produced by a simple change of sign for one of the coefficients in the Walsh decomposition of f, provided the global optimum remains unchanged. However, if the global optimum changes as well, ρ_{fd} may also change, whereas it makes no difference to η. To this extent, fitness-distance correlation may be a more discriminating measure than epistasis variance.

8.7 Reference classes

Naudts and Kallel [180] have considered the question of problem difficulty from a different angle. They discuss several potential measures of difficulty, including some of those mentioned above, and relate each measure to a class of functions for which the measure attains an extreme value out of all those possible. The value of epistasis variance, for example, will attain its minimum (zero) for functions that are *first-order* or fully *separable*—those of the form

$$f(\boldsymbol{x}) = \sum_{i=1}^{\ell} g_i(x_i)$$

for functions $g_i(\cdot)$ that map string values to $\mathbb{R}$. If these individual functions are linear, we often describe $f(\boldsymbol{x})$ as linear, but in [180] this term is given a more restricted meaning— specifically, that the fitness value is a linear function of the Hamming distance of the string from the global optimum. Naudts and Kallel also use a different normalization from that we use for η, which leads to an ambiguity in assessing *constant* or flat functions (a degenerate subset of first-order functions). According to their version of epistasis variance, there is no way to distinguish them. However, using Equation (8.3), constant functions lead to an indeterminate value rather than to zero. Thus for η as defined in this chapter, the reference class must exclude the case of constant functions.

In principle, using the approach of [180] makes it possible to compute the value of a stated measure for a specified function, and compare it to that of a member of the reference class. However, as [180] makes clear, interpreting the resulting ratio as a measure of difficulty should be done with caution. As well as epistasis variance, epistasis correlation and fitness-distance correlation, they discuss other concepts. Of these, we shall describe one that can be taken as characteristic of their approach.

Definition 8.2 *The* sitewise optimization *algorithm takes an arbitrary string and searches independently for the best allele for each gene. A sitewise optimizable function is then one for which this algorithm delivers the global optimum for all input strings.*

This algorithm was originally developed by Wilson [313] in an attempt to distinguish functions that were 'GA-easy' but difficult for hill-climbers. For binary strings, it simply means testing a bit-flip at each locus. Such **swo** functions generalize the notion of first-order functions, which are clearly a special case.

Suppose we wish to use this idea to propose a measure of difficulty that is relevant for members of the **swo** class. In order to define such a measure, we use the Hamming distance metric to describe the topological relationships between vectors. The *width* of a set of points is defined to be the average Hamming distance between its members. Now imagine applying the **swo** algorithm to a set of points; the **swo** measure is the ratio of the widths of the sets of points obtained 'before' and 'after' the algorithm is applied. If the function has a unique global optimum, this measure will of course be zero.

It is also shown in [180] that **swo** functions are a proper subset of 'GA-easy' functions (if we define these in terms of schemata and deception), and of a restricted version[5] of steepest-ascent (as defined in Chapter 1) optimizable functions, but that neither of these wider classes of functions includes the other.

From the ED perspective, first-order functions are those whose Walsh decomposition has interaction terms that are all zero, while GA-easy functions admit interactions, but only if they are benign. What makes for benign or malign interactions is also discussed in [180], along with a number of other classes of functions and related measures, and some results concerning their relationships.

---

[5]The restriction lies in the fact that once the value at a locus has been fixed by the steepest ascent algorithm, it cannot be modified again. This means that the algorithm is guaranteed to halt in no more than ℓ iterations.

8.8 General Discussion

The previous sections deal with some of the flaws inherent in methods aimed at predicting problem difficulty, and attempts to correct some of the misapprehensions that still exist in the literature. The link with statistical methods also demonstrates that we need not always try to invent theoretical tools from scratch: other disciplines may have some interesting light to throw upon our subject.

8.8.1 Implications for schema theory

Turning first to the experimental design connection, it is instructive to compare and contrast the processes typically used in ED with Holland's schema-processing arguments. Firstly, an assumption is made concerning the nature of the function under investigation: in particular, what is believed about the maximum level of interactions. This is then used to design a set of experiments that confound higher-order interactions with those of lower order so that the effects associated with the latter can all be estimated without the aliasing problem. Then these estimates are used in order to determine those factors which are the most important, and those levels of each factor which achieve the best average performance. Finally, these are combined together in order to provide the best overall setting of factors and levels. Further confirmatory tests can then be carried out if desired, and new hypotheses can be explored based on the results of these tests.

Does this sound familiar? According to schema-processing theory, many of these processes are in fact going on *implicitly* when we run a GA: implicit schema averages play the part of the explicit estimates of the effects, fitness-proportional selection does the job of sorting out which effects are most important, and recombination assembles the fitter building blocks into new chromosomes that implicitly test new hypotheses.

Some experimental work reported in [225] compared several statistical methods for effectiveness in finding high-quality solutions to a design problem. At each stage a small population was evaluated and the alleles that appeared to produce the worst performance were eliminated. (This was a sort of 'stripped-down' GA: the process was explicit and deterministic, rather than the implicit, stochastic approach used by a normal GA.) Different elimination criteria were examined: using mean performance (i.e. schema averages) was markedly inferior to using schema *maxima* as the criterion. Analysis showed that deception did exist in this problem, so we would expect schema averages to provide ambiguous results. A 'normal' (steady-state) GA was also run on the same problem, and it proved far more robust and less easy to deceive—which suggests that there is indeed rather more to GAs than schema processing.

8.8.2 Can prediction work?

The whole *idea* of predictive measures raises two questions of a more fundamental nature, however. Firstly, can we ever know for certain how difficult an optimization problem is, without exploring the whole Universe? The arguments of this chapter suggest that simple measures such as η or ρ_{fd} just cannot guarantee to tell us how difficult a problem is on their own, even supposing we can estimate their values reliably. We should not be surprised that *different* problems have the same values of these measures, but the situation is worse in that, for any given problem, there are many *closely-related* problems whose Walsh coefficients differ only in their signs, and epistasis variance cannot distinguish them. In fact, by simply changing the signs of the Walsh coefficients [228], we can generate problems with very different characteristics in terms of numbers of local optima (with respect to some specified operator), sizes of attraction basins etc., and yet their η and ρ_{fd} values are identical. In the most general case, therefore, we need to know all about the Walsh decomposition in order to know how difficult the problem is. So measuring the difficulty of the problem is as hard as solving it.

Secondly (as a corollary to the above), even if we could find out, would it be worth the effort? The presumed motivation for trying to measure problem difficulty is to enable us to use an appropriate combination of algorithm, representation and operators. Culberson puts this dilemma neatly:

> *[We are] placed in the unfortunate position of having to find a representation, operators and parameter settings to make a poorly understood system solve a poorly understood problem....[we might] be better served concentrating on the problem itself.* [42]

Of course, it might be objected, this is something of a worst-case scenario. 'Real-world' problems need not be so bad: we could focus on problems of 'bounded epistasis'—where we understand this term to mean that there are no interaction effects beyond a pre-defined order. The simplest case is that of a first-order function, where there are no interactions of any order.

English's information-theoretic viewpoint is clearly relevant at this point. As discussed in Chapter 4, he has shown that for almost all functions, learning is hard—and it is learning that we are attempting when we aim to predict the difficulty of solving an optimization problem. Given that optimization

8.8. GENERAL DISCUSSION

is relatively *easy* (in English's terms), it appears that we have exchanged one problem for a more difficult one. However, we are bound to wonder whether there is something to be learned for problems that are restricted in some way, and using the Walsh/ED representation provides an obvious way of defining restricted classes of functions.

Culberson's adversary argument [42], as discussed in Chapter 4, may also be useful here (*pace* English's strictures against it [69]). In the general case, Culberson's adversary is completely unconstrained and can pass back any value whatsoever. It is clear that in such cases it can also defeat any attempt to predict the difficulty of the problem on the basis of previous evaluations. At the other extreme, if the function is first-order, the adversary is highly constrained in its 'evaluation' function, and cannot report an evaluation that is inconsistent with its past history.

Suppose then that we have a case of bounded epistasis. What sort of algorithm would give us the *best* performance? Consider an algorithm $\mathcal{E}$ that proceeds as follows: it repeatedly chooses a string to pass to the adversary, using ED principles and related assumptions about the nature of the function. On completing a certain number of trials $\mathcal{E}$ is confident it has exact estimates of all the effects so that it can predict the optimum (which it may not necessarily have seen yet). If $\mathcal{E}$ is correct in its assumptions, the adversary is 'boxed in': any future evaluations have to concur with the algorithm's predictions, and it cannot hide the optimum from view. Note, however, that if $\mathcal{E}$'s prediction is based on insufficient data, some at least of the estimated effects will be incorrect, and the adversary can still falsify its prediction, even if the assumptions as to the *nature* of the function are valid. Is it possible to calculate how many trials $\mathcal{E}$ needs before its prediction cannot be gainsaid?

In the case of a first-order function, ED theory gives some guidelines. For a string of length ℓ, a 'Resolution-III' (R-III) design[6] can guarantee to compute all the main effects (i.e., order 1 Walsh coefficients) free of confounding. Such a design for a 3-bit problem is shown in Table 8.4.

With the half-fraction specified by the R-III design, the logical estimate of α_1 is

$$\hat{\alpha}_1 = \frac{(f_1 + f_7) - (f_2 + f_4)}{4}$$

---

[6] A Resolution-r design is one in which no m-factor interaction is confounded with an interaction of fewer than n factors, where $m + n = r$. Thus a R-III design does not confound main effects ($m = 1$) with each other, but main effects are confounded with 2-factor interactions. In general, the existence of such designs for all ℓ, r is still an open question, and their construction is complicated, depending on group theoretic concepts and the properties of orthogonal arrays.

Table 8.4: A R-III design for $\ell = 3$. This should be compared with the half-fraction in Table 8.2. In this case, all main effects are distinguishable from each other.

fitness value	String Universe	R-III
f_0	0 0 0	
f_1	0 0 1	0 0 1
f_2	0 1 0	0 1 0
f_3	0 1 1	
f_4	1 0 0	1 0 0
f_5	1 0 1	
f_6	1 1 0	
f_7	1 1 1	1 1 1

which is really the value of

$$\alpha_1 + (\beta\gamma)_{11}.$$

Thus in this case α_1 and $(\beta\gamma)_{11}$ are aliased; it is readily checked that a similar relationship is true for all main effects. By way of comparison, the confounding seen earlier in Table 8.2 is less neat: what we had there was a linear combination of two main effects α_1 and β_1 masquerading as simple estimates of α_1. The important point for Table 8.4 is that for a linear function, the interaction terms are zero, so $\hat{\alpha}_1$ is a valid estimate of the true value α_1. In a similar way, both β_1 and γ_1 can be estimated clear of main-effect aliases from this design, so that we can predict exactly the value of any string. Of course, we want to locate the optimum without needing to evaluate every possible string, but assigning the best alleles at each locus is easy in this case, since there are no interactions.

For the case of R-III, then, it can be seen that *if* we choose the 'right' subset of 4 strings to send to the adversary, it has no more room for manoeuvre. Furthermore, there is no subset of 3 strings that would allow us to produce valid estimates of all 3 main effects. Algorithm $\mathcal{E}$ really is the best algorithm we could use in this case, as it is bound to 'defeat' the adversary in the minimum number of evaluations. In general, for a string of length ℓ, choosing a R-III design entails a maximum of 2^t evaluations, where $t = \lceil \log_2(\ell + 1) \rceil$, and finding such a design is fairly easy. Thus the growth in the number of evaluations is essentially linear in the string length. In the more general case, this merely puts a lower bound on the number of evalua-

8.8. GENERAL DISCUSSION

tions: if we take random samples instead of orthogonal ones, the adversary will usually be able to keep fooling the algorithm for longer, although how much longer is an open question.

This sounds promising, but as so often, once we move away from the linear case, things are considerably bleaker. As soon as we consider epistatic order-2 functions (no interactions above 2 factors), it is already hard. Firstly, $\mathcal{E}$ needs a R-V design in order to compute the effects, which is rather more difficult to construct. Secondly, even if the effects have all been computed correctly, we still have not solved the initial problem. As we pointed out earlier, in order to predict the optimum, $\mathcal{E}$ now has to decide how to choose the correct alleles at each of the appropriate gene loci, and even if we limit the interactions to second-order, this is effectively a MAXSAT problem, which is NP-hard [118, 178].

It is true that an evolutionary algorithm such as a genetic algorithm is different from our idealized algorithm $\mathcal{E}$: GAs use random sampling instead of orthogonal arrays, and don't calculate the values of the effects explicitly (although schema-processing arguments suggest that it is happening implicitly). However, it can scarcely be argued that these changes are to the advantage of a GA, as any such implicit investigation of epistasis is likely to be subject to greater error in its estimates of the effects.

Finally, in practice the *assumption* of bounded epistasis might still be incorrect. The above analysis has only asked how much data we need if the assumptions hold. In reality, we could use $\mathcal{E}$ with the prescribed set of function evaluations and thus compute all the effects up to our (presumed) maximum order, solve the satisfiability problem, and then find out that our assumptions were all wrong. The adversary could simply make use of the aliased effects—the ones we (falsely) assumed were negligible—in order to defeat us.

8.8.3 Conclusion

The above arguments sound extremely gloomy and may suggest that not merely the idea of investigating epistasis, but indeed the whole question of finding near-optimal solutions to optimization problems is inherently doomed. From the information-theoretic perspective of English, as reviewed in Chapter 4, predicting the performance of an algorithm is really more like learning than optimization—which means swapping an 'easy' problem for a 'hard' one.

However, the lessons of years of development and practice of experimental design methods (as well as GAs) suggest that the assumption of bounded epistasis may well be a good approximation for real-world problems. Essentially, in many practical problems interaction effects seem to 'die away' quite quickly as the order of interaction grows. Thus methods based on estimating the size of the effects and extrapolation using some assumption about the level of epistasis has worked very well in practice. The 'problem' may well be mainly epistemological and not ontological—that is, although our results may be entirely satisfactory, we nevertheless want to *know* whether they are or not, and indeed to *know that we know*!

Essentially, we see that *really* to know is usually equivalent to complete knowledge of the Universe. The idea that there is a 'magic bullet'—an algorithm that is the best for our problem—is a nice conceit, but the effort it would need to find it is almost certainly more than we can afford. Perhaps we should drop the search for a magic bullet and use a scatter gun instead—just try to solve the problem by whatever techniques we have available. After all, even if we think we have found the magic bullet, in reality we can never be sure that we haven't fooled ourselves!

8.9 Bibliographic Notes

Yuval Davidor [45, 46] was the first to try to characterise the epistasis of a specific function, but it was not entirely successful. Christine Wright and Colin Reeves made the connection to statistics and placed these ideas on a firmer basis [217, 218]. Various interesting and relevant facts about Walsh functions can be found in papers by Rob Heckendorn, Soraya Rana and Darrell Whitley [114, 118, 115, 117, 116]. The idea of fitness-distance correlation was initially introduced to GAs by Bernard Manderick and co-workers [157], and extended by Terry Jones and Stephanie Forrest [137]. Further work by these and other workers is summarized in [223]. The review paper by Bart Naudts and Leila Kallel [180] is also a source of much useful information.

Exercises

1. Using Yates' algorithm (see Exercise 6 of Chapter 3) or otherwise, find the Walsh coefficients for the functions $F_1, \ldots, F_4$ as displayed below, and hence determine their epistasis variance.

string	F_1	F_2	F_3	F_4
0 0 0	10	10	20	0
0 0 1	50	50	30	90
0 1 0	30	70	10	70
0 1 1	70	30	40	100
1 0 0	20	40	0	60
1 0 1	60	80	30	90
1 1 0	40	60	10	70
1 1 1	80	20	100	80

2. (continuation) Compare the coefficients for F_1 and F_2. What do you observe? [Hint: think of Gray codes.] What are the implications of this observation?

3. (continuation) Consider the *signs* of the coefficients for F_3 and F_4. What is the implication of this observation for the difficulty of optimizing the two functions?

4. Consider the fraction in the table below:

fitness value	String Universe	fraction
f_0	0 0 0	
f_1	0 0 1	
f_2	0 1 0	0 1 0
f_3	0 1 1	
f_4	1 0 0	1 0 0
f_5	1 0 1	1 0 1
f_6	1 1 0	1 1 0
f_7	1 1 1	

   Verify that the 'obvious' estimate

   $$\hat{\alpha}_1 = \frac{3f_5 - f_2 - f_4 - f_6}{6}$$

actually estimates the linear combination

$$\frac{6\alpha_1 - 4\beta_1 - 2(\alpha\beta)_{11} + 2\gamma_1 + 4(\alpha\gamma)_{11} - 2(\beta\gamma)_{11} - 4(\alpha\beta\gamma)_{111}}{3}$$

[Hint: Show that the coefficients in the combination c satisfy the equation $c^T = b^T \widetilde{W}$, where b contains the weights used to find the estimate, and $\widetilde{W}$ is the Walsh transform matrix as defined in Chapter 3.]

5. Calculate the epistasis correlation for function F_4 of Exercise 1 and verify the relationship of Proposition 8.2.

6. Calculate the value of ρ_{fd} for each of the functions $F_1, \ldots, F_4$ of Exercise 1, and investigate the existence of a Δ value as defined in Proposition 8.3.

Chapter 9

Landscapes

Up to this point, we have nearly always assumed (at least implicitly) that the topological relationship between the vectors $x \in \mathcal{X}$ is best described by the simple Hamming distance. However, while this seems an obvious choice for neighbourhood search algorithms, and for aspects of GAs that relate to mutation, it is much less obvious that it is relevant for recombination. In addition, for representations that depend on permutations, for example, whatever the choice of algorithm, Hamming distance may make little sense. This raises the whole question of what we actually mean when we talk about a 'fitness landscape'.

9.1 Introduction

Jones [135] was perhaps the first to stress the importance of realising that the 'landscape' observed for a particular function is an artefact of the algorithm used or, more particularly, the neighbourhood induced by the operators it employs. It is thus unfortunate that it is still common to find references in the literature to the fitness *landscape* when it is simply the fitness *function* that is intended. Sometimes, as in the case of the so-called 'NK-landscapes' [144], the Hamming neighbourhood is assumed, almost by default, but this is not *prima facie* appropriate to GAs, for example, although it is relevant to hill-climbing algorithms based on neighbourhood search.[1]

With the idea of a fitness landscape comes the idea that there are also many local optima or false peaks, in which a search algorithm may become trapped without finding the global optimum. In continuous optimization,

---

[1] For those unfamiliar with the idea, a description of the procedure for implementing an NK-function is given in Appendix A.

notions of continuity and concepts associated with the differential calculus enable us to characterize quite precisely what we mean by a landscape, and to define the idea of an optimum. It is also convenient that our own experiences of hill-climbing in a 3-dimensional world gives us analogies to ridges, valleys, basins, watersheds etc., which help us to build an intuitive picture of what is needed for a successful search, even though the search spaces that are of interest often have dimensions many orders of magnitude higher than 3.

However, in the continuous case, the landscape is determined only by the fitness function, and the ingenuity needed to find a global optimum consists in trying to match a technique to this single landscape. There is a major difference when we come to discrete optimization. Indeed, we really should not even use the term 'landscape' until we define the topological relationships of the points in the search space $\mathcal{X}$. In the continuous case, the detailed topology can be altered by a transformation of the underlying variables, which would correspond to a change of representation in discrete problems. However, independently of the representation, the topology for discrete search spaces is also specified by the search algorithm with which we choose to search it.

9.1.1 An example

The idea that we can have a 'fitness landscape' without reference to the search algorithm is surprisingly still believed, and the contrary notion is even considered by some people to be controversial. This is odd, as it is easily demonstrated that the significant features of a landscape vary with changes in search operators and strategy. In practice, one of the most commonly used search methods for COPs is *neighbourhood search* (NS), which we have already described in Chapter 1. This idea is also at the root of modern 'metaheuristics' such as simulated annealing (SA) and tabu search (TS)—as well as being much more involved in the methodology of GAs than is sometimes realized.

A *neighbourhood* is generated by using an operator that transforms a given vector x into a new vector x'. It can be thought of as a function

$$N : \mathcal{X} \mapsto 2^{\mathcal{X}}$$

For example, if the solution is represented by a binary vector (as it often is for GAs), a simple neighbourhood might consist of all vectors obtainable by 'flipping' one of the bits. The 'bit flip' (BF) neighbourhood of $\mathbf{0} = (0\,0\,0\,0\,0)$,

9.1. INTRODUCTION

for example, would be the set

$$N(0) = \{(1\,0\,0\,0\,0), (0\,1\,0\,0\,0), (0\,0\,1\,0\,0), (0\,0\,0\,1\,0), (0\,0\,0\,0\,1)\}$$

Consider the problem of maximizing a simple cubic function

$$f(x) = x^3 - 60x^2 + 900x + 100$$

where the solution x is required to be an integer in the range $[0, 31]$. We have already met this problem in Chapter 2. Regarding x as a continuous variable, we have a smooth unimodal function with a single maximum at $x = 10$—as is easily found by calculus—and since the solution is already an integer, this is undoubtedly the most efficient way of solving the problem.

However, suppose we chose instead to represent x by a binary vector $\boldsymbol{x}$ of length 5 as described in Chapter 2. By decoding this binary vector as an integer it is possible to evaluate f, and we could then use NS, for example, to search over the binary hypercube for the global optimum using some form of hill-climbing strategy.

This discrete optimization problem turns out to have 4 optima (3 of them local) when the BF neighbourhood is used. If a 'steepest ascent' strategy is assumed (i.e., the *best* neighbour of a given vector is identified before a move is made) the local optima and their basins of attraction are as shown in Table 9.1. On the other hand, if a 'first improvement' strategy is used (where the first change that leads uphill is accepted without ascertaining if a still better one exists), the basins of attraction are as shown in Table 9.2. We can think of this as occurring because some points lie in a 'col' between two (or possibly more) peaks, and which route the search takes in the case of first improvement is dictated by the order in which the upward directions are encountered.

That this order can be rather important is illustrated in Table 9.3, where the order of searching the components of the vector is made in the reverse direction (right-to-left) whereas in Table 9.2 it was left-to-right. The optima are still the same, of course,[2] but the basins of attraction are very different.

This simple problem clearly shows two important points. Firstly, by using BF with this binary representation, we have created local optima that did not exist in the integer version of the problem. Secondly, although the optima are still the same, the chances of reaching a particular optimum can be seriously affected by a change in hill-climbing strategy. However, we can

---
[2] Note that the last cycle of first improvement, after an optimum is attained, is always identical to that of steepest ascent.

Table 9.1: Local optima and basins of attraction for steepest ascent with the BF operator in the case of a simple cubic function. The bracketed figures are the fitnesses of each local optimum.

Local optimum	01010 (4100)	01100 (3988)	00111 (3803)	10000 (3236)
Basin	00000	00100	00110	10000
	00001	01100	00111	10001
	00010	11100	10110	10010
	00011		10111	10011
	00101			10100
	01000			
	01001			
	01010			
	01011			
	01101			
	01110			
	01111			
	10101			
	11000			
	11001			
	11010			
	11011			
	11101			
	11110			
	11111			

9.1. INTRODUCTION

Table 9.2: Local optima and basins of attraction for first improvement (forward search) using the BF operator.

Local optimum	0 1 0 1 0	0 1 1 0 0	0 0 1 1 1	1 0 0 0 0
	(4100)	(3988)	(3803)	(3236)
Basin	0 0 1 0 1	0 0 1 0 0	0 0 1 1 1	0 0 0 0 0
	0 0 1 1 0	0 1 0 0 0	0 1 1 1 1	0 0 0 0 1
	0 1 0 0 1	0 1 1 0 0	1 0 1 1 1	0 0 0 1 0
	0 1 0 1 0	1 0 1 0 0	1 1 1 1 1	0 0 0 1 1
	0 1 0 1 1	1 1 0 0 0		1 0 0 0 0
	0 1 1 0 1	1 1 1 0 0		1 0 0 0 1
	0 1 1 1 0			1 0 0 1 0
	1 0 1 0 1			1 0 0 1 1
	1 0 1 1 0			
	1 1 0 0 1			
	1 1 0 1 0			
	1 1 0 1 1			
	1 1 1 0 1			
	1 1 1 1 0			

also demonstrate very clearly our major contention: that the landscape is altered by choice of neighbourhood, for the bit flip operator is not the only mechanism for generating neighbours. An alternative neighbourhood could be defined as follows: for $k = 1, \ldots, 5$, flip bits $\{k, \ldots, 5\}$. For example, the neighbourhood of $\mathbf{0} = (0\,0\,0\,0\,0)$, would now be

$$N(\mathbf{0}) = \{(1\,1\,1\,1\,1), (0\,1\,1\,1\,1), (0\,0\,1\,1\,1), (0\,0\,0\,1\,1), (0\,0\,0\,0\,1)\}.$$

We shall call this the CX operator and its neighbourhood the CX neighbourhood. It creates a very different landscape: there is now only a single global optimum (0 1 0 1 0), i.e., *every* vector is in the one basin of attraction. (And obviously, it no longer matters whether the strategy is steepest ascent or first improvement, nor in which order the neighbours are evaluated.) Thus we have clearly changed the landscape by changing the operator. *It is not merely the choice of a binary representation that generates the landscape—the search operator needs to be specified as well.*

Incidentally, there are two interesting facts about the CX operator. Firstly, it is closely related to the one-point crossover operator frequently used in genetic algorithms. It was for that reason it was named [122] the complementary crossover or CX operator. Secondly, if the 32 vectors in the search

Table 9.3: Local optima and basins of attraction for first improvement (reverse search) using the BF operator.

Local optimum	01010 (4100)	01100 (3988)	00111 (3803)	10000 (3236)
Basin	01000	01100	00000	10000
	01001	01101	00001	10001
	01010	01110	00010	10010
	01011	01111	00011	10011
			00100	10100
			00101	10101
			00110	10110
			00111	10111
				11000
				11001
				11010
				11011
				11100
				11101
				11110
				11111

space are interpreted as elements of a *Gray* code, it is easy to show that the neighbours of a point in Gray-coded space under BF are identical to those in the original binary-coded space under CX. This is illustrated in Table 9.4, and is an example of an *isomorphism* of landscapes.

Table 9.4: Illustration of the isomorphic relationship between (BF, Gray) and (CX, binary) landscapes. The neighbours of the point (00000) are listed together with their respective decoded integer values.

	BF-Neighbours	Gray-Integer
00000	00001	1
	00010	3
	00100	7
	01000	15
	10000	31
	CX-Neighbours	binary-Integer
00000	00001	1
	00011	3
	00111	7
	01111	15
	11111	31

In a pioneering paper on this topic, Culberson [41] used something rather similar to the CX operator to demonstrate that the 'GA-easy' *Onemax* function (see Appendix A) does not in fact have an easy 'crossover landscape'. We shall consider this example at greater length later in the chapter.

9.2 Mathematical Characterization

The general idea of a fitness landscape is usually credited to the population geneticist Sewall Wright [318], although a similar notion had already been discussed by Haldane [110]. The metaphor became extremely popular, being employed in many biological texts as a metaphor for Darwinian evolution. However, Wright's use of this metaphor is actually rather ambiguous, as Provine [194] has pointed out, and its application to evolution is somewhat questionable. The modern concept in biology commenced with a paper by

Eigen [66], and concerns mainly the properties of RNA sequences.[3] Nevertheless, it is also a helpful metaphor in the area of optimization, although it is only recently that a proper mathematical characterization of the idea has been made. The most comprehensive treatment is that by Stadler [269], and space allows only a relatively brief discussion of all the ideas that Stadler and his colleagues have developed.

We can define a landscape $\mathcal{L}$ for the function f as a triple $\mathcal{L} = (\mathcal{X}, f, d)$ where d denotes a distance measure $d : \mathcal{X} \times \mathcal{X} \to \mathbb{R}^+ \cup \{\infty\}$ for which it is required that

$$\left. \begin{array}{l} d(s,t) \geq 0 \\ d(s,t) = 0 \Leftrightarrow s = t \\ d(s,u) \leq d(s,t) + d(t,u) \end{array} \right\} \quad \forall\, s, t, u \in \mathcal{X}.$$

Note that we have not specified the representation explicitly (for example, binary or Gray code): this is assumed to be implied by the descriptions of $\mathcal{X}$ and f. We have also decided, for the sake of simplicity, to ignore questions of search strategy and other matters in the definition of a landscape, unlike some more comprehensive definitions, such as that of Jones [135], for example, which try to take into account probabilistic effects.

This definition says nothing either about how the distance measure arises. In fact, for many cases a 'canonical' distance measure can be defined. Often, this is symmetric, i.e. $d(s,t) = d(t,s)\ \forall\, s, t \in \mathcal{X}$, so that d also defines a *metric* on $\mathcal{X}$. This is clearly a nice property, although it is not essential.

9.2.1 Neighbourhood structure

What we have called a canonical distance measure is typically related to the neighbourhood structure. Every solution $t \in N(s)$ can be reached directly from s by a *move*—a single application of an operator ω to a vector s in order to transform it into a vector t. The canonical distance measure d_ω is that induced by ω whereby

$$t \in N(s) \Leftrightarrow d_\omega(s,t) = 1.$$

The distance between non-neighbours is defined as the length of the shortest path between them (if one exists). As the landscape is fundamentally a

---

[3]RNA molecules are made up of long sequences using just 4 bases, commonly denoted by the letters A, C, G and U. The RNA 'sequence space' is therefore enormously vast, and the landscape of interest is that induced by point mutations, i.e., substitutions of one 'letter' by another.

9.2. MATHEMATICAL CHARACTERIZATION

graph, standard graph algorithms for finding shortest paths [181, 266] can easily be applied if required.

For example, if $\mathcal{X}$ is the binary hypercube $\{0,1\}^\ell$, the bit flip (BF) operator acting on the binary vector $x \in \{0,1\}^\ell$ can be defined as

$$\phi(k): \{0,1\}^\ell \to \{0,1\}^\ell \quad \begin{cases} x'_k = 1 - x_k \\ x'_i = x_i & \text{if } i \neq k \end{cases}$$

where $x' \in N(x)$. It is clear that the distance metric induced by ϕ is the Hamming distance $d_H(x, x')$, as defined in Chapter 8. Thus we could refer to this landscape as a Hamming landscape (with reference to its distance measure), or as the BF landscape (with reference to its operator). Similarly, we can define the CX operator as

$$\gamma(k): \{0,1\}^\ell \to \{0,1\}^\ell \quad \begin{cases} x'_i = 1 - x_i & \text{for } i \geq k \\ x'_i = x_i & \text{otherwise} \end{cases}$$

Clearly, it would be quite legitimate to define a landscape as $\mathcal{L} = (\mathcal{X}, f, \omega)$, but the specification with respect to distance is more general, albeit most distance measures used in practice are actually induced by an operator. However, there are occasions where operator-independent measures are useful. There is a discussion on this point in [224].

9.2.2 Local optima

We can now give a formal statement of a fundamental property of fitness landscapes: for a landscape $\mathcal{L} = (\mathcal{X}, f, d)$, a vector $s \in \mathcal{X}$ is *locally optimal* if

$$f(s) > f(t) \ \forall \ t \in N(s).$$

In some cases, we might wish to relax the criterion by replacing $>$ by $\geq$. This would allow for the existence of 'flat-topped' peaks (or *mesas*, as they are called in [135]). For the sake of simplicity, we shall ignore such questions here. Landscapes that have only one local (and thus also global) optimum are commonly called *unimodal*, while landscapes with more than one local optimum are said to be *multimodal*.

The number of local optima in a landscape clearly has some bearing on the difficulty of finding the global optimum. That this is relevant has often been assumed in research on problem difficulty (for example, in [311]), and the general intuition is sound. However, it is not the only indicator: the size of the basins of attraction of the various optima is also an important influence. We could imagine, for example, a problem instance having a large number of local optima that were almost 'needles in a haystack', while the basin of attraction of the global optimum occupies a large fraction of the space. Such a problem would not be difficult.

9.2.3 Graph representation

Neighbourhood structures are clearly just another way of defining a graph Γ, which can be described by its $(n \times n)$ *adjacency matrix* A. The elements of A are given by $a_{ij} = 1$ if the indices i and j represent neighbouring vectors, and $a_{ij} = 0$ otherwise. For example, the graph induced by the bit flip ϕ on binary vectors of length 3 has adjacency matrix

$$A_\phi = \begin{bmatrix} 0 & 1 & 1 & 0 & 1 & 0 & 0 & 0 \\ 1 & 0 & 0 & 1 & 0 & 1 & 0 & 0 \\ 1 & 0 & 0 & 1 & 0 & 0 & 1 & 0 \\ 0 & 1 & 1 & 0 & 0 & 0 & 0 & 1 \\ 1 & 0 & 0 & 0 & 0 & 1 & 1 & 0 \\ 0 & 1 & 0 & 0 & 1 & 0 & 0 & 1 \\ 0 & 0 & 1 & 0 & 1 & 0 & 0 & 1 \\ 0 & 0 & 0 & 1 & 0 & 1 & 1 & 0 \end{bmatrix}$$

where the vectors are indexed in the usual binary-coded integer order (i.e., $(0\,0\,0), (0\,0\,1)$ etc). By way of contrast, the adjacency matrix for the CX operator is

$$A_\gamma = \begin{bmatrix} 0 & 1 & 0 & 1 & 0 & 0 & 0 & 1 \\ 1 & 0 & 1 & 0 & 0 & 0 & 1 & 0 \\ 0 & 1 & 0 & 1 & 0 & 1 & 0 & 0 \\ 1 & 0 & 1 & 0 & 1 & 0 & 0 & 0 \\ 0 & 0 & 0 & 1 & 0 & 1 & 0 & 1 \\ 0 & 0 & 1 & 0 & 1 & 0 & 1 & 0 \\ 0 & 1 & 0 & 0 & 0 & 1 & 0 & 1 \\ 1 & 0 & 0 & 0 & 1 & 0 & 1 & 0 \end{bmatrix}$$

It is simply demonstrated that permuting the rows and columns of A_ϕ so that they are in the order $0, 1, 3, 2, 7, 6, 5, 4$ reproduces the matrix A_γ— another way of demonstrating the isomorphism mentioned earlier. In other words,

$$P^{-1} A_\phi P = A_\gamma$$

where P is the associated permutation matrix of the binary-Gray transformation. It is also follows from standard matrix theory that the eigenvalues and eigenvectors are the same.

9.2.4 Laplacian matrix

The *graph Laplacian* Δ may be defined as

$$\Delta = A - D$$

where D is a diagonal matrix such that d_{ii} is the degree of vertex i. Usually, these matrices are vertex-regular and $\forall i \ \forall x$, $d_{ii} = k = |N(x)|$, so that

$$\Delta = A - kI$$

This notion recalls that of a Laplacian operator in the continuous domain; the effect of this matrix, applied as an operator at the point s to the fitness function f is

$$\Delta f(s) = \sum_{t \in N(s)} (f(t) - f(s))$$

so it functions as a kind of differencing operator. In particular, $\Delta f(s)/k$ is the average difference in fitness between the vector s and its neighbours. Grover [109] has shown that the landscapes of several COPs satisfy an equation of the form

$$\Delta f(s) + \frac{C}{m} f(s) = 0$$

for all points s, where C is a problem-specific constant and m is the size of the problem instance. From this it can be deduced that *all* local optima are better than the mean ($\bar{f}$) of all points on the landscape. (In other words, all minima are lower, all maxima are higher than the mean.) Furthermore, it can also be shown that under mild conditions on the nature of the fitness function, the time taken by NS to find a local optimum in a maximization problem is $\mathcal{O}(m \log_2[f_{max}/\bar{f}])$ where f_{max} is the fitness of a global maximum. (Modifications for minimization problems are straightforward.)

9.2.5 Random Walks

Once we have a notion of adjacency, we can envisage a search algorithm as traversing a path through the landscape. As we have argued, different heuristics induce different landscapes, but even on the same landscape, performance can depend markedly on the strategy that is followed. (Recall the example discussed in Section 9.1.1, where 'first improving' paths led rather infrequently to the global optimum.) Such strategies can be stochastic, deterministic, or they can try to pick up information from their local environment. An algorithm that assumes nothing else about the landscape in order to follow such a path would be random search, and the path followed would be a *random walk*. In this case, we can describe its progress in terms of Markov chains, with a transition matrix

$$T = AD^{-1}$$

(Of course, for a vertex-regular graph, this simplifies to $T = A/k$.) Weinberger [300] showed that certain quantities obtained in the course of such a random walk can be useful indicators of the nature of a landscape. If the fitness of the point visited at time t is denoted by f_t, we can estimate the *autocorrelation function* (usually abbreviated to *acf*) of the landscape during a random walk of length T as

$$r_j = \frac{\sum_{t=1}^{T-j} (f_t - \bar{f})(f_{t+j} - \bar{f})}{\sum_{t=1}^{T} (f_t - \bar{f})^2}.$$

Here $\bar{f}$ is of course the mean fitness of the T points visited, and j is known as the *lag*. The concept of autocorrelation is an important one in time series analysis (see, for example, [28]), but its interpretation in the context of landscapes is interesting.

For 'smooth' landscapes, and at small lags (i.e., for points that are close together), the *acf* is likely to be close to 1 since neighbours are likely to have similar fitness values. However, as the lag increases the correlations will diminish. 'Rugged' landscapes are informally those where close points can nevertheless have completely unrelated fitnesses, and so the *acf* will be close to zero at all lags. Landscapes for which the *acf* has significant negative values are conceptually possible, but they would have to be rather strange.

A related quantity is the *correlation length* of the landscape, usually denoted by τ. Classical time series analysis [28] can be used to show that the standard error of the estimate r_j is approximately $1/\sqrt{T}$, so that there is only approximately 5% probability that $|r_j|$ could exceed $2/\sqrt{T}$ by chance. Values of r_j less than this value can be assumed to be zero. The correlation length τ is then the last j for which r_j is non-zero:

$$\tau = j : |r_{j+1}| < 2/\sqrt{T} \wedge \{|r_k| > 2/\sqrt{T} \quad \forall k \leq j\}$$

9.2.6 Graph eigensystem

In the usual way, we can define eigenvalues and eigenvectors of the matrices associated with a graph. The set of eigenvalues is called the *spectrum* of the

9.2. MATHEMATICAL CHARACTERIZATION

graph. The spectrum of the Laplacian is

$$\begin{pmatrix} \mu_0 & \mu_1 & \cdots & \mu_{n-1} \end{pmatrix}$$

where μ_i is the ith eigenvalue, ranked in (weakly) ascending order. There is of course a close relationship between the spectrum of the Laplacian and the spectrum of $\mathbf{A}$: if the graph is regular and connected, $\mu_i = k - \mu_i^A$, where μ_i^A is the ith eigenvalue of $\mathbf{A}$, ranked in (weakly) *descending* order. Normally, the graphs in which we are interested are connected, in which case the smallest eigenvalue $\mu_0 = 0$, with multiplicity 1. From the eigenvectors $\{\varphi_i\}$ corresponding to μ_i, f can be expanded as

$$f(\mathbf{x}) = \sum_i a_i \varphi_i(\mathbf{x}).$$

Stadler and Wagner [270] call this a 'Fourier expansion'. Grover's results (quoted above) showed that some COPs—the TSP, for example—have the form

$$f(\mathbf{x}) = c + \varphi(\mathbf{x}),$$

where c is a constant and $\varphi(\mathbf{x})$ is an eigenvector corresponding to a non-zero eigenvalue. Such landscapes are termed *elementary*; it is clear that the Fourier expansion is a way of decomposing a landscape by a superposition of elementary landscapes.

Usually, the eigenvalues are not simple, and the expansion can be further partitioned by the *distinct* eigenvalues of $\mathbf{\Delta}$, denoted by $\{\lambda_p\}$. The corresponding values

$$\beta_p = \sum_{i : \mu_i = \lambda_p} |a_i|^2$$

form the *amplitude spectrum*, which expresses the relative importance of different components of the landscape. In fact, these values are usually normalized (denoted in what follows by β_p') by dividing by $\sum_{i>0} |a_i|^2$.

The eigenvalues can also be related to the *acf*. It can be shown [269] that

$$r_j = \sum_{p>0} \beta_p' (1 - \lambda_p/k)^j,$$

while the correlation length is given by

$$\tau = \sum_{j=0}^{\infty} r_j = k \sum_{p>0} \frac{\beta_p'}{\lambda_p}.$$

Ideally, such mathematical characterizations could be used to aid our understanding of the important features of a landscape, and so help us to exploit them in designing search strategies. But beyond Grover's rather general results above, it is possible to carry out further analytical studies only for small graphs, or graphs with a special structure. Angel and Zissimopolous [8] provide an analysis of several well-known COPs in terms of their autocorrelation functions.

In the case of Hamming landscapes it is possible to find analytical results for the graph spectrum, which show that the eigenvectors are thinly disguised versions of the familiar Walsh functions. For the case of recombinative operators the problem is rather more complicated, and necessitates the use of 'P-structures' [270]. The latter are essentially generalizations of graphs in which the mapping is from pairs of 'parents' (x, y) to the set of possible strings that can be generated by their recombination. However, it can be shown that for some 'recombination landscapes' (such as that arising from the use of uniform crossover) the eigenvectors are once more the Walsh functions. Whether this is generally true in the case of 1X or 2X, for example, is not known, but Stadler and Wagner show that it is so in some cases, and conjecture that the result is general. In view of the close relationship between the BF and CX landscapes as demonstrated above, it would not be surprising if this is a general phenomenon. However, to obtain these results, some assumptions have to be made—such as a uniform distribution of parents—that are unlikely to be true in a specific finite realization of a genetic search.

In the case of the BF or Hamming landscape, the distinct eigenvalues correspond to sets of Walsh coefficients of different orders, and the (normalised) amplitude spectrum is exactly the set of components used in calculating the 'epistasis variance' η that we have discussed in Chapter 8. For the cubic function of Section 9.1.1 above, the components of variance for the different orders of Walsh coefficients can be shown to be $(0.387, 0.512, 0.101, 0, 0)$ respectively, with $\eta = 0.613$. It is clear that the interactions predominate and in this case indicate the relatively poor performance of the BF hill-climber.

Of course, the eigenvalues and eigenvectors are exactly the same for the CX landscape of this function, and the set of Walsh coefficients in the Fourier decomposition is also the same. However, the effect of the permutation inherent in the mapping from the BF landscape to the CX landscape is to re-label some of the vertices of the graph, and hence some of the Walsh coefficients. Thus some coefficients that previously referred to linear effects now refer to interactions, and vice-versa. Taking the cubic function as an example again, the components of variance or amplitude spectrum becomes

9.2. MATHEMATICAL CHARACTERIZATION

(0.771, 0.174, 0.044, 0.011, 0.000), with $\eta = 0.229$. We see that the linear effects now predominate, and this is consistent with the results we obtained from the CX hill-climber.

9.2.7 Recombination landscapes

If we look at the 'recombination landscapes' derived from Stadler's and Wagner's P-structures [270], we find that once again the Walsh coefficients are obtained, but labelled in yet another way. The coefficients in the BF and CX landscapes are grouped according to the number of 1s in their binary- and Gray-coded index representations respectively. However, in a recombination landscape—such as that generated by 1-point crossover (1X)—it is the *separation* between the outermost 1-bits that defines the groupings. Table 9.5 shows the groupings for a 4-bit problem.

Table 9.5: Illustration of the different groupings of the Walsh coefficients associated with the BF CX and 1X recombination landscapes. The groups are labelled 0,1,2,3 and 4 in each case.

Index	binary coding	BF	CX	1X	Index	binary coding	BF	CX	1X
0	0000	0	0	0	8	1000	1	2	1
1	0001	1	1	1	9	1001	2	3	4
2	0010	1	2	1	10	1010	2	4	3
3	0011	2	1	2	11	1011	3	3	4
4	0100	1	2	1	12	1100	2	2	2
5	0101	2	3	3	13	1101	3	3	4
6	0110	2	2	2	14	1110	3	2	3
7	0111	3	1	3	15	1111	4	1	4

Several things can be seen from this table: firstly, (as already explained), the coefficients in the CX landscape are simply a re-labelling of those in the BF landscape. Secondly, the linear Walsh coefficients (and hence the values of η) are the same in both the BF and the 1X landscapes. Thirdly, the coefficients in the recombination landscape do not form a natural grouping in terms of interactions, and consequently the different components of variance for the recombination landscape do not have a simple interpretation as due to interactions of a particular order. However, the fact that the linear components are the same goes a little way towards assuaging the qualms raised in Chapter 8 concerning the usefulness of epistasis variance as a measure of the difficulty of a given problem instance for a GA.

Nevertheless, the amplitude spectrum only gives a rough guide to the difficulty or otherwise of a given landscape, and how a GA 'creates' its landscape in any particular case is not something that can be stated easily. However, we can use analogy and intuition to gain some sort of understanding, even if it is difficult in the current state of theory to make strong numerical predictions. In the next two sections, we shall discuss some special cases.

9.3 *Onemax* Landscapes

It is instructive to work through what happens with a simple function—in this case, the *Onemax* function (see Appendix A). Firstly, as it is a separable function, the only non-zero Walsh coefficients are the linear ones:

$$w_0 = \ell/2 \tag{9.1}$$

$$w_j = \begin{cases} -1/2 & \text{for } j = 2^k; \ k = 0, \ldots, \ell - 1 \\ 0 & \text{otherwise} \end{cases} \tag{9.2}$$

(For those preferring the ED 'labels', these non-zero coefficients are μ, α_0, β_0 etc.) Thus, the components of variance are exactly the same for a hill-climber in the Hamming landscape and for a GA using 1X, so on a simplistic reading of the meaning of the amplitude spectrum, *Onemax* should be equally easy for both.

In practice, it is evident that this is not so: a BF hill-climber will find the optimal string for *Onemax* rather quickly: no more than ℓ function evaluations will ever be needed if we use a next ascent strategy. In fact, this somewhat overstates the computational requirements, since the whole function does not actually need to be evaluated (merely the effect of a single bit change), and on average $\ell/2$ bits will be correct even before the search starts. However, a GA without mutation[4] finds *Onemax* relatively hard to solve.

This might seem an odd statement to make: it is often claimed that *Onemax* is 'GA-easy'—even that it is the easiest function for a GA. Here we are comparing its *relative* performance, and that is much worse than a hill-climber in the Hamming landscape. However, a CX hill-climber is

---
[4]We ignore mutation here, as this complicates the analysis of the effect of crossover. With mutation only, to a first-order approximation, a GA is essentially exploring the Hamming landscape. However, while it proceeds well at first, it slows down as it tries to fix the last few 1-bits and is much less efficient than a deterministic hill-climber.

9.3. ONEMAX LANDSCAPES

much worse than either. Figure 9.1 shows some typical performance curves for these different approaches, omitting the hill-climber in the Hamming landscape, since it would be scarcely visible on the same scale as the others.

The number of function evaluations clearly grows somewhat faster than the Hamming landscape hill-climber's linear performance—something like $\mathcal{O}(\ell \log \ell)$ seems a plausible growth rate. So although the amplitude spectra are the same, there is a clear difference in the *computational* effectiveness of a GA and a hill-climber. This raises an interesting question: how exactly does the GA (or more specifically, recombination) solve the *Onemax* problem? Figure 9.1 also shows that while the GA solves both versions of the *Onemax* problem more easily than a CX hill-climber, it is less efficient on the Gray-encoded version than the standard one. This finding also seem to need explanation.

9.3.1 How recombination solves *Onemax*

Firstly, we need to consider the CX landscape of the *Onemax* function. This can be shown (using an argument first put forward by Culberson [41]; see also Exercises 2, 3 and 4 at the end of this chapter) to have a very large number of local optima—a number that increases exponentially with ℓ. This is reflected by the fact that a CX hill-climber needs an exponentially increasing amount of computation, as shown in Figure 9.1. We can easily calculate the amplitude spectrum by using the Gray code isomorphism that we mentioned above. The Walsh coefficients of the CX landscape are the same as in the Hamming landscape, but with their labels permuted. In the Hamming landscape, the indices of the non-zero coefficients are all powers of 2; in other words, the binary encoding of these indices have a single 1 and $\ell - 1$ zeros. Their Gray code transformations will be the values

$$0\ldots0\,0\,1, \quad 0\ldots0\,1\,1, \quad 0\ldots1\,1\,0, \quad \text{etc.}$$

The first corresponds to a main effect, but the rest correspond to indices of second-order interaction effects. It thus follows from Equation (8.5) that the epistasis variance is high even for moderate string lengths, at

$$\frac{(\ell-1)(1/2)^2}{\ell(1/2)^2} = 1 - \frac{1}{\ell},$$

while the amplitude spectrum is $(1/\ell, (\ell-1)/\ell, 0, \ldots, 0)$. So the landscape analysis would indeed predict the poor performance of a CX hill-climber.

The recombination landscape is clearly identical to the CX landscape in the event that the selected parents are complementary pairs—not that

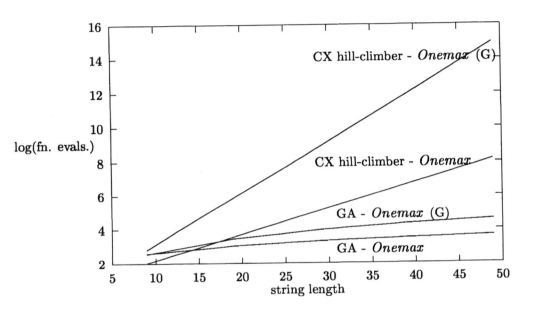

Figure 9.1: Performance of various algorithms for *Onemax* and Gray-encoded *Onemax* functions. The vertical axis measures the number of function evaluations needed to find the optimum (averaged over 30 independent trials), and is on a logarithmic scale. The hill-climber used a first improvement strategy in the CX landscape (which is, as already noted, identical to the BF landscape of the Gray-encoded *Onemax*). The GA was generational with binary tournament selection, one-point crossover, no mutation and a population of 200. In all cases, the search was restarted if it converged to a point that was not the global optimum.

9.3. ONEMAX LANDSCAPES

this is likely in practice. However, the recombination landscape described by Stadler's and Wagner's P-structures is essentially based on all possible pairs of parents, uniformly distributed—again, not what we find in practice. Starting from a random initial population, the first generation parents are on average $\ell/2$ bits different, and for large ℓ the chance of a significant deviation from this number is small. The P-structure approach is useful in providing a general picture, but in its basic form it fails to distinguish a recombinative GA from a Hamming landscape hill-climber. Stadler *et al.* [271] have recently proposed a methodology for dealing with non-uniformly distributed populations. Their analysis is highly technical, but the basic result is that the effect of non-uniformly distributed populations is to modify the *acf*, in a way which suggests that the landscape will appear to be 'smoothed out'. However, while this is demonstrable as a general tendency, what we would like is some idea of what happens at a localized level in a specific case.

In other words, we need some idea of the *realized* crossover landscape for a particular population—i.e., the set of points that are actually reachable, and their networks of neighbourhood relations. We can think of the crossover landscape as being somewhere 'between' the CX landscape and the Hamming landscape. The first case corresponds to the recombination of parents that are maximally distant, while the second corresponds to parents that are minimally distant—having only two bits different[5]. Of course, this is rather vague, and the argument that follows will be qualitative and intuitive in nature. Nevertheless, it receives some support from the theoretical approach of [271].

9.3.2 The dual effect of recombination

The impact of recombination on the CX landscape is twofold. For the CX landscape the distance $d_\gamma(\boldsymbol{x}, \boldsymbol{y}) < \infty$, $\forall\ \boldsymbol{x}, \boldsymbol{y} \in \{0, 1\}^\ell$, which means that it is *connected*. However, the crossover landscape is composed of *disconnected* fragments defined, when the parents are $\boldsymbol{x}$ and $\boldsymbol{y}$, by the vector $\boldsymbol{d} = \boldsymbol{x} \oplus \boldsymbol{y}$ (where $\oplus$ is addition modulo 2). Any element of $\boldsymbol{d}$ that is zero corresponds to a common allele in the parents, and so reduces the search space. For example, as a result of crossing the binary strings (0 1 0) and (1 0 0), we

---

[5]Having one bit different might seem to be minimal, but in this case recombination could not generate a child that was different from one of its parents, and so no move on the landscape would actually take place.

would have $d = (1\,1\,0)$, so that we would be searching the space $(*\,*\,0)$.[6]

Thus the original landscape has been fragmented, and recombination restricts the current search to one of the fragments. Elsewhere [122] this has been called the primary effect of recombination. The size of the fragments is determined by the Hamming distance between the parents. Since the average Hamming distance between parents shrinks during search under the influence of selection, the search becomes restricted to smaller and smaller fragments.

But that is not all that recombination can do: it also effects a particular type of linear transformation of the CX landscape with the property that neighbourhoods (and therefore potentially local optima) are not preserved. In particular, the neighbourhood of a point in a fragment of (say) the 1X landscape is, in general, not a subset of its neighbourhood in the fully connected CX landscape, as is illustrated in Figure 9.2. Thus we should not imagine that the fragments are merely disconnected components of the original landscape as would be the case with bit-flips. If we applied a similar linear transformation in the Hamming landscape, it is easy to see that the neighbourhoods would be subsets of the original larger one.

This is the secondary effect of recombination, which is important for understanding what happens to local optima. By means of this secondary effect, the search can escape from local optima that are present in the underlying CX landscape. While this is not the only candidate for explaining how a GA circumvents such traps, Höhn and Reeves [122] described some experiments to verify that this is a plausible explanation.

Furthermore, we are also entitled to ask if the chosen *mechanism* of crossover is actually important. After all, we could use *any* vector d to accomplish the primary effect of creating a fragmented CX landscape. Jones [136] was the first to point out the relevance of this question, in slightly different terms from the way we have posed it here—as separating the 'idea' of crossover from its 'mechanics'. His concern was to distinguish the effect of crossover from that of large-scale or macro-mutation.

He suggested that in testing the relevance of crossover in any application, we should consider crossover in the context of 'headless chicken' selection. For the latter, only one parent is chosen from the current population, while the other is a random string. This is obviously the same as choosing a randomly generated binary vector d in our exposition above, so we can interpret headless chicken selection and crossover as forming *different* fragments of the

---

[6]Of course, this is also a schema; again we see the close relationship between recombination and schemata that we remarked upon in Chapter 3.

9.3. ONEMAX LANDSCAPES

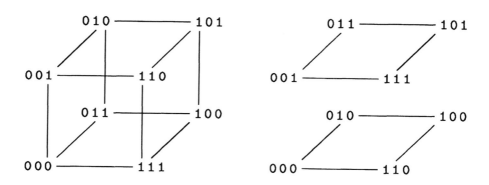

Figure 9.2: Two effects of recombination: while the neighbourhood graph of the CX landscape (left) is connected, the 1X landscape (right) consists of disconnected fragments. In this case, $d = (1\,1\,0)$, so that the search is restricted to the lower fragment. Furthermore (the secondary effect), the neighbours of a particular point are different: the string $(0\,0\,0)$, for example, is now adjacent to two points that were previously at a distance of 2.

CX landscape—but different in what way?

In the 'headless chicken' experiments the distance between the selected parents is on average about $\ell/2$, and varies little throughout the search. In the traditional parent selection scheme, by contrast, the distance between parents is a function of the population composition, and it tends to decrease as the population converges.

Thus, the distance between parents determines the size of the CX landscape fragments, and as this distance decreases, it also changes the nature of the landscape. For example, if the parents are complementary, i.e. if they are at maximal Hamming distance, the 1X landscape is identical to the CX landscape. At the other extreme, when the parents are at Hamming distance 2 from each other, the crossover landscape is a fragment of the Hamming landscape.

Consequently, adjusting the distance between parents makes the crossover landscapes 'look' more like (fragments of) the CX landscape, or more like (fragments of) the Hamming landscape. Because selection is directional in its effect, this means that a GA can solve problems that are difficult in the CX landscape but easy in the Hamming landscape (such as the *Onemax* function). Conversely, we would predict that the GA will be less able to

Table 9.6: Two 'Gray-encoded' Onemax functions for the case $\ell = 3$. The function f is the standard Onemax, f_{GB} and f_{BG} are as defined in the text.

x	f	f_{GB}	f_{BG}
0 0 0	0	0	0
0 0 1	1	1	1
0 1 0	1	2	2
0 1 1	2	1	1
1 0 0	1	3	2
1 0 1	2	2	3
1 1 0	2	1	2
1 1 1	3	2	1

solve problems that are difficult in the Hamming landscape.

This prediction can easily be tested by trying to solve a Gray-encoded *Onemax* function. By Gray-encoded in this case we mean a function $f_{GB}(x)$, say, where x is interpreted as a Gray-coded string, which must be decoded to its standard binary interpretation before applying *Onemax*. (In other words, $f_{GB}(x)$ is the composition $f(GB(x))$, where $GB(x)$ is the Gray-to-binary encoding as discussed in Chapter 4.) Note that we could generate another 'Gray-encoded' function by interpreting x as a standard binary string, Gray-coding it, and applying *Onemax* to the result, i.e., a function $f_{BG}(x) = f(BG(x))$. The two cases are illustrated in Table 9.6

The f_{GB} function corresponds to the isomorphism discussed above in Section 9.1.1. Therefore, because the CX landscape of *Onemax* is difficult, the Gray-encoded version will be hard in the Hamming landscape, having many local optima that create severe problems for a hill-climber. (It can also be shown [122] that the CX landscape of Gray-encoded *Onemax* faces similar difficulties.)

How does a GA fare? The results for this case were also plotted in Figure 9.1. It is clear that the GA does much better than the CX hill-climber for Gray-encoded *Onemax*, but less well than it does for standard *Onemax*—just as predicted. As the population converges the GA generates landscape fragments that are similar to the Hamming landscape. In contrast, if we use the 'headless chicken' GA, all the landscape fragments generated lie somewhere between the CX landscape and the Hamming landscape, and so we would expect to find no difference between the two *Onemax* versions. This is exactly what happens, as can be seen from Figure 9.3, where the two

9.4. LONG PATH LANDSCAPES

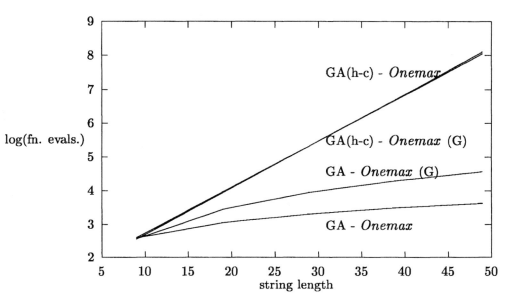

Figure 9.3: Performance of the 'headless chicken' GA for *Onemax* and Gray-encoded *Onemax*. For comparison, the results of the original GA are also included. As before, the vertical axis measures the number of function evaluations needed (averaged over 30 independent trials), and is on a logarithmic scale.

relevant performance curves are indistinguishable.

Thus it is clear that solving the *Onemax* problem requires both the primary and the secondary effect of crossover to be working in harmony. This is something that is not always appreciated. When research has shown that 'crossover does not work' for a given problem, we should really ask the further question: why? Which of the effects is responsible for this observation? Perhaps pairwise mating (which establishes the primary effect) is actually a good idea, but our choice of operator is a poor one. We shall return to this subject a little later in Section 9.6.

9.4 Long Path Landscapes

From a hill-climbing perspective, multiple optima create a serious problem, but multimodality is not the only cause of difficulties. Horn *et al.*

[130] pointed out that even landscapes with a single global optimum will be hard for a Hamming hill-climber if the path is excessively long, and suggested several ways of constructing functions that will exhibit such behaviour. Moreover, their experiments suggested that long path problems are cases where GAs are likely to outperform hill-climbers. However, it was not clear whether such problems are intrinsically hard or easy for a GA, or how a GA actually goes about solving such cases.

9.4.1 Paths

We can formally define a path as follows:

Definition 9.1 *A path of length k in the landscape $\mathcal{L} = (\mathcal{X}, f, d)$ is a sequence $x_0, \cdots, x_k$ of points in the search space $\mathcal{X}$ such that $d(x_i, x_{i+1}) = 1$ for $i = 0, \ldots, k-1$. Furthermore, a path $x_0, \cdots, x_k$ is strictly monotonic increasing if $f(x_{i+1}) > f(x_i)$ for $i = 0, \ldots, k-1$.*

9.4.2 The *Root2path*

The simplest long path is the ℓ-dimensional *Root2path* in the landscape $(\{0,1\}^\ell, f, d_H)$. It is constructed recursively, as described in detail in Appendix A. As an example, the 5-dimensional *Root2path* is given by the 11-point sequence

$$00000 \quad 00001 \quad 00101 \quad 00111 \quad 00110 \quad 10110$$
$$11110 \quad 11111 \quad 11101 \quad 11001 \quad 11000$$

It can be seen that, apart from the 'bridge point' (number 6), the path segments (points 1-5 and 7-11) characteristically fold back on themselves, which gives such functions the alternative name of 'snake-in-a-box' [280]. It is simple to make this path a strictly monotonic increasing one, by defining f to be a monotonic function of the position of a point on the path. Points not on the path are given values by the *Zeromax* function,[7] so that they lead towards the initial point on the path, which is **0**—although it is possible that we encounter a path-point before we reach **0**. The resulting composite function is called the *Extended Root2path*. If we choose to add a suitable offset (such as ℓ) to path points, they will be fitter than any off-path points, and so each point on the path has just one better Hamming neighbour: the next point on the path. The global optimum is the final point on the path, $x = (1\,1\,0\,0\,\cdots\,0\,0\,0)$, but it will take a hill-climber a long time to

---
[7]*Zeromax* is just like *Onemax*, but we wish to maximize the number of 0s.

9.4. LONG PATH LANDSCAPES

arrive. (Note, however, that if we could flip two bits at a time, we could take shortcuts that would substantially reduce the path length. In fact, the 'earlier' we are in the path, the further we could jump by a double bit flip.)

Höhn and Reeves [123] replicated the GA experiments of [130], with typical results as shown in Table 9.7. In this case, a generational GA was used with population 200, binary tournament selection and 1X at a crossover rate $\chi = 1$. As in [130], no mutation was used, and if the optimum was not found when the population converged, the GA was restarted with a new random population.

ℓ	path length n	Function evaluations				Restarts
		MIN	MAX	MEAN	SEMEAN	MEAN
9	47	200	800	441	29	0.0
19	1535	1062	16144	5505	770	2.2
29	49151	3866	106814	29583	4577	11.3
39	1572863	2808	246990	73673	10124	23.4
49	50331647	11378	1442224	349207	61449	95.4

Table 9.7: Performance of a standard generational GA on the *extended Root2path* problem. The number of function evaluations required to solve the problem grows exponentially but at a slower rate than the size of the path. A linear regression analysis suggests a growth rate of $\approx 1.17^{\ell}$. These data were taken from 30 independent runs.

Although the GA finds the optimum while evaluating only a fraction of the Universe, the number of evaluations needed still appears to grow exponentially. Thus, one could hardly call this problem 'easy' for a GA, so dashing the hopes of [130] that long paths could serve as a means of distinguishing GAs from other search algorithms. However, what the GA actually does in solving the *Extended Root2path* is rather interesting. If we consider only points on the path, there is some useful structure [123], akin to the 'Royal Road' functions of [169]. The schemata

$$
\begin{array}{c}
1\,1\,*\,*\,*\,*\cdots*\,*\,* \\
\,\,0\,0\,*\,*\cdots*\,*\,* \\
\,\,*\,*\,0\,0\cdots*\,*\,* \\
\,\,*\,*\,*\,*\cdots 0\,0\,* \\
\,\,*\,*\,*\,*\cdots*\,*\,0
\end{array}
$$

define 'building blocks'—partial solutions for the optimum of the *Root2path*.

Further, of the points on the path, only 1/3 (approximately) are bridge points; the rest are all members of these schemata, so there should be great potential for constructing the optimum according to building block theory. Unfortunately, most of the space is outside the path, but the *Zeromax* function is supposed to do the job of finding the path, and according to the ideas presented in Section 9.3, this is not too difficult. So is this a case with a 'royal road', one where the GA works at least partially by means of building blocks, once some points on the path are discovered? This would be noteworthy, given the current scepticism concerning building block theory, as discussed in Chapter 3.

Sadly the answer seems still to be in the negative. The evidence presented in [123] demonstrates that the GA more often than not simply stumbles upon the *Root2path* optimum while looking for that of *Zeromax*. As Table 9.8 shows, the parents of the optimal string are rarely on the path themselves. Only for small values of ℓ would a 'building-block' explanation seem plausible. Since the optimum of the *Root2path* is only Hamming distance 2 from that of *Zeromax*, it is quite easy to see how it could be found 'by accident'.

	First parent			Second parent		
ℓ	Off-path	Non-bridge	Bridge	Off-path	Non-bridge	Bridge
9	0.31	0.54	0.15	0.85	0.15	0.00
19	0.80	0.13	0.07	0.93	0.07	0.00
29	0.87	0.10	0.03	0.83	0.17	0.00
39	1.00	0.00	0.00	0.90	0.10	0.00
49	0.90	0.10	0.00	0.93	0.07	0.00

Table 9.8: Parental origin of the optimal string found by the GA, showing the fraction of parents in each class, averaged over 30 experiments. Clearly, when solving the *extended Root2path*, the majority of optimal solutions are found by crossover between elements that lie off the path. The share of non-bridge points is too small to attribute the success of the GA to the 'royal road' in the *Root2path*.

9.5 Local Optima and Schemata

The *Onemax* analysis has suggested that the Hamming landscape may be important for successful GAs. Jones [135] laid stress on the fact that the

9.5. LOCAL OPTIMA AND SCHEMATA

GA's landscape is not the same as the Hamming landscape, yet it appears that investigating the latter may not be a complete waste of time. One of the characteristics of NS methods on a landscape is the occurrence of multiple local optima—points to which the search converges. According to the dynamical systems model discussed in Chapter 6, GAs also have points to which they converge in the infinite population case. How do these compare with NS-optima? At least in the case of mutation, we might well expect them to coincide—after all, to first-order, a mutation landscape 'looks like' the Hamming landscape.

Rana [208] argued that the convergence points, or *attractors*, of a GA for binary strings must also be local optima of the Hamming landscape. The argument, based on the statistical law of large numbers, is that eventually the effect of selection and mutation will ensure that a Hamming local optimum is attained. (Of course, with positive mutation rates, perturbations from this point will occur, and in the limit, the fixed points of such a GA will not be corners of the simplex. Nevertheless, we might reasonably expect that most of the population will be concentrated at these points.)

Rana only considered mutation effects, although she did also explore the relationship between local optima and schema processing issues. She calculated the 'static ϕ' measure (discussed in Chapter 8) for Hamming landscape local optima, and compared it to their basin sizes for several NK-landscapes and MAXSAT problems. The rankings obtained from the two approaches tended to coincide quite well.

This still leaves open the question—what about crossover? If we restrict ourselves to crossover and selection (no mutation), then potentially *every* point is an attractor: just start the process with multiple copies of the same string (recall from Chapter 6 that we call this a uniform population) and it will stay there. However, not all fixed points are stable—suppose just one copy of a better Hamming neighbour is introduced into an otherwise uniform population: the only strings that can be generated will be these two neighbouring strings, so it is clear that (at least for infinite populations) the better one will eventually take over. Confining our definition of an attractor to those points that *are* asymptotically stable, it is clear that any such string with a Hamming neighbour of greater fitness cannot be a fixed point for a crossover-only GA. Thus every GA attractor must be a local optimum of the Hamming landscape, although the converse is not true. (It is easy to find counter-examples.) The above argument is intuitive, but it can be made rigorous by taking advantage of the dynamical systems model described in Chapter 6. The key theorem is the following:

Theorem 9.1 (Vose-Wright [291]) *For a crossover-only GA, the spectrum of $d\mathcal{G}_{e_k}$ is given by:*

$$spec(d\mathcal{G}_{e_k}) = \left\{ \frac{f_{i\oplus k}}{f_k} \sum_u (\chi_u + \chi_{\bar{u}})[u \otimes i = 0] : i = 1, 2, \ldots, n-1 \right\} \bigcup \{0\}$$

This needs a little unpacking: χ_u is the probability of applying the crossover mask u, whose complement is the mask $\bar{u}$; the (bitwise) operators $\otimes$ and $\oplus$ are AND and XOR respectively. Also note that for notational simplicity the statement of this theorem does not distinguish between vectors and indices: for example, k is used both as an index (e.g., in f_k) and as a vector (e.g., in $i \oplus k$).

The point of this is that the status of a particular uniform population vector e_k can be determined by looking at the set of eigenvalues—the spectrum. If we consider masks defining the 'standard' types of crossover (1X, UX etc.), it is possible to derive fairly straightforward conditions for particular uniform populations to be stable 'attractors'. It turns out [232, 291] that all the attractors must be Hamming local optima[8]. Thus, whether we see the landscape from the point of view of mutation or crossover, it seems that the Hamming local optima play an important rôle.

9.5.1 How many attractors/optima?

The problem with analyses such as we have described is that we can often only verify useful results after the event: we need to know the Universe. Moreover, in the case of landscapes, we need to know the status of every point: is it a local optimum (for NS landscapes) or an attractor (for GAs)? Clearly, in practice we need some way of assessing landscape properties without this exhaustive knowledge. Reeves [229, 230, 231] and (independently) Kallel and Garnier [140] have shown that it is possible to obtain estimates of the number of attractors by statistical analysis of the results of repeated random restarts using the principle of maximum likelihood. In the case of isotropic landscapes—those where the attractors are distributed in such a way that the basins of attraction are identical in size—the estimates have been shown experimentally to be quite good. More recent work by Eremeev and Reeves [71] has shown that *non-parametric* statistical methods give much better estimates for non-isotropic landscapes.

---

[8]Strictly speaking, this is only so if the conjecture that fixed points in the finite population case are uniform populations is true.

The *number* of attractors is of course not the only factor in problem difficulty—the relative *size* of the basins of attraction is also clearly of importance. A problem instance might have few optima, but if the global optimum is effectively a 'needle-in-a-haystack', finding it can be very difficult for any algorithm. Even in cases where there is quasi-gradient information for a specific landscape of a problem, effective search is not guaranteed, as we saw earlier in the case of long path functions. Nevertheless, in practice we often do find that the number of optima has a significant impact on the efficiency of a search, and it pays to have a landscape with fewer attractors.

9.6 Path Tracing GAs

The detailed 'topography' of a landscape may also have an important bearing on effective search. If we had a completely isotropic landscape, where all attractors were identical in fitness and basin size, optimization would be easy—in the sense that any attractor would do. On the other hand, if the basins were identical in size, but attractor fitness varied substantially but (in a 'geographical' sense) randomly, repeated random restarts of NS would most likely be as good as anything.

The interesting cases would be where attractors have basin sizes that were correlated to their fitnesses, and where the location of one attractor provides information on the locations of others, so that 'intelligent' search methods have something to work with. In many cases, perhaps a little surprisingly, this appears to be very close to what we do see.

There have been several empirical studies of landscapes, starting with Lin's experiments [152] on the travelling salesman problem—although the term 'landscape' was not then defined, or even commonly used. One of the recurring features of these studies has been given the name of the 'big valley' conjecture. This is seen in the NK-landscapes formulated by Kauffman [144], and it also appears in many examples of COPs, such as the TSP [152, 25], graph partitioning [163], and flowshop scheduling [224].

These studies have repeatedly found that, on average, local optima are very much closer to the global optimum than are randomly chosen points, and closer to each other than random points would be. That is, the distribution of local optima is *not* isotropic; rather, they appear to be clustered in a 'central massif', or—if we are minimizing—a 'big valley'. This can be visualized by plotting a scatter graph of fitness against distance to the global optimum. Moreover, if the basins of attraction of each local optimum are explored, size is quite highly correlated with quality: the better the local

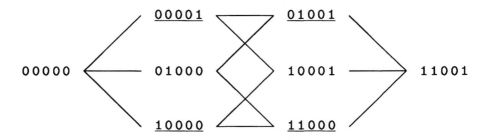

Figure 9.4: The diagram shows the set of paths that could be traced between the parents 0 0 0 0 0 and 1 1 0 0 1. Only those intermediate vectors indicated by underlining can be generated by one-point crossover, but all can be generated by uniform crossover.

optimum, the larger is its basin of attraction.

Of course, there is no guarantee that this property holds in any particular case, but it provides an explanation for the success of 'perturbation' methods [134, 158, 320] which currently appear to be the best available for the TSP. It is also tacitly assumed by such methods as simulated annealing and tabu search, which would lose a great deal of their potency if local optima were really distributed isotropically.

It also suggests a starting point for the development of new ways of implementing evolutionary algorithms. Reeves and Yamada [222, 226] proposed that we could exploit the big valley by interpreting crossover as 'path-tracing'. If we consider the case of crossover of vectors in $\{0,1\}^\ell$, it is easily seen that any 'child' produced from two 'parents' will lie on a path that leads from one parent to another. Figure 9.4 illustrates this fact.

In an earlier paper [214] such points were described as 'intermediate vectors', in the sense that they are intermediate points on the Hamming landscape. Thus crossover is re-interpreted as finding a point lying 'between' 2 parents in some landscape for which we hope the big valley conjecture is true. This interpretation is in full agreement with the 'dual effect' of recombination as discussed earlier in Section 9.3.2. The common elements are always inherited, but the choice of the remaining ones can perhaps be done in a more 'sensible' way than a random guess. In other landscapes the distance measure may be more complicated, but the principle remains. This 'path-tracing crossover' was implemented for both the makespan and

9.7. BIBLIOGRAPHIC NOTES

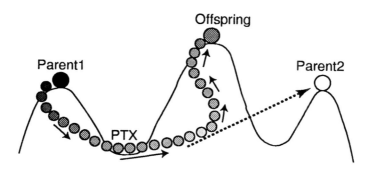

Figure 9.5: Path tracing crossover combined with local search: a path is traced from one parent in the direction of the other. In the 'middle' of the path, solutions may be found that are not in the basins of attraction of the parents. A local search can then exploit this new starting point by climbing to the top of a hill (or the bottom of a valley, if it is a minimization problem)—a new local optimum. The acronym PTX signifies 'path-tracing crossover'.

the flowsum versions of the flowshop sequencing problem; Figure 9.5 shows in a 2-dimensional diagram the idea behind it, while full details can be found in [222, 226].

The results of this approach (the details can be found in [222]) were far superior to those of methods based on simpler crossover operators, such as PMX and others discussed in Chapter 2, and demonstrate that making use of a theoretical perspective on a problem can be of considerable practical effect. This approach has also recently been applied to multi-constrained knapsack problems by Levenhagen et al. [150], who confirmed the need for a 'big valley' structure in order to gain some benefit from this approach.

9.7 Bibliographic Notes

The major contributors to the theory of landscapes have been Peter Stadler and his co-workers at the University of Vienna and at the Santa Fe Institute. Stadler's seminal work [269] is essential (if challenging) reading for a solid mathematical foundation. A more recent review paper [234] discusses some of the latest research in this still flourishing area. The Vienna University website
http://www.tbi.univie.ac.at/~studla/publications.html

contains these and many related papers by Stadler and other major contributors such as Günter Wagner and Edward Weinberger. The Santa Fe Institute is also an outstanding source for material on this topic: http://www.santafe.edu/sfi/publications/working-papers.html.

The notion of NK-functions, which has become rather popular in GA research, can be found in the thought-provoking writings of Stuart Kauffman [144, 145]. These were generalised to NKP-'landscapes' independently by Lee Altenberg [7] and Rob Heckendorn [115], and later further generalised to ℓ, θ-functions by Colin Reeves [228]. The close connection of NK-'landscapes' to MAXSAT problems was first pointed out by Heckendorn, Soraya Rana and Darrell Whitley [118]. Alden Wright and co-workers [316] established the nature of the computational complexity of NK-functions.

Another important work is the thesis of Terry Jones[135], which is rather more accessible to the general reader than Stadler's *magnum opus*. Jones also has two related papers [136, 137], the latter one with Stephanie Forrest. Christian Höhn and Colin Reeves applied a qualitative reading of the landscape formulation to some standard problems [122, 123]. Rana's thesis [208] contains an in-depth study of the properties of local optima.

Empirical work on the nature of landscapes has also been carried out in the field of operational research [25]; contributions have been made in the context of evolutionary algorithms by Peter Merz and Bernd Freisleben [163, 164], and by Colin Reeves and Takeshi Yamada [222, 224].

A review covering some of the material in this chapter can be found in the book edited by Leila Kallel, Bart Naudts and Alex Rogers [142], which also includes Kallel's work [140] on estimating the number of local optima. Independent research by Colin Reeves of a similar nature is reported in [229], and further developments by Reeves and Anton Eremeev can be found in [71].

9.7. BIBLIOGRAPHIC NOTES

Exercises

1. Determine the amplitude spectra of the functions $F_1, \ldots, F_4$ (Exercise 1 of Chapter 8) for BF, CX and 1X. (Refer to Table 9.5 for guidance.)

2. Show that the conditions for a local optimum in the CX landscape of the *Onemax* function are

$$\sum_{i=k}^{\ell} x_i > \frac{\ell - k + 1}{2} \quad \text{for } k = 1, \ldots, \ell$$

   Hence deduce that the last two bits of a local optimum must be 1s, and that more than half the bits must be 1s. Show that there are 3 local optima for $\ell = 4$ and 6 for $\ell = 5$.

3. (continuation) Consider the case of ℓ an even number, and denote the number of local optima by n_ℓ. Suppose we prefix the strings with the bits (0 1), (1 0) or (1 1) respectively. Hence show that

$$n_{\ell+2} \geq 3 n_\ell.$$

   Starting from the case $\ell = 4$, use this result to derive an expression for a lower bound on the growth of the number of optima with ℓ.

4. For the case ℓ an odd number, classify local optima into types A and B according as the first bit is 0 or 1, and denote the number of such local optima for strings of length ℓ by a_ℓ, b_ℓ respectively. Use a similar argument to show that

$$\begin{aligned} a_{\ell+2} &\geq a_\ell + b_\ell \\ b_{\ell+2} &\geq a_\ell + 2 b_\ell. \end{aligned}$$

   Starting from $\ell = 5$, derive an expression for a lower bound on the growth of the number of optima with ℓ. [Hint: It is the case (1 0) that needs special treatment. Also, finding the lower bound on growth is most easily handled by using generating functions.]

5. Construct a long path problem where at least 3 bits would have to be flipped to get from one segment of the path to an adjacent one.

Chapter 10

Summary

The previous chapters have provided considerable detail on the state of GA theory at the start of the 21st century. We have traced the development of the early ideas on schemata and building blocks, on implicit parallelism and the two-armed bandit; we have described some of the failings of these ideas, and we have discussed several new perspectives that are relevant in attempting to understand better how genetic algorithms really work. However, we are aware that it is not uncommon for readers to read the first and last chapters of a book before going in deeper, so we shall take the opportunity in this last chapter of summarising the key points of our earlier discussions. This will mean some repetition, but perhaps that is no bad thing.

10.1 The Demise of the Schema Theorem

The theory of genetic algorithms has come a long way since the publication of Holland's book [124]. Most of the so-called 'theoretical' explanation for GA behaviour based on ideas in that book has been shown to have flaws. In its place, we now have a formal mathematical framework, as described in Chapter 6, in which properties of GAs can be rigorously proven. In fact, the original schema theorem has its proper place within this framework, but it is now known that one cannot appeal naively to the schema theorem to explain genetic algorithm behaviour. In particular, the *building block hypothesis* has had to be substantially revised and the notion of *implicit parallelism* debunked [297].

The problem comes from a misinterpretation of the effect of selection and mixing on schemata. Holland's emphasis was on the effects of selection in increasing the proportion of schemata with above average fitness. Crossover and mutation were seen as having a modifying destructive effect. There are two problems here. Firstly, it has become clear that the *constructive* effects of mixing are of fundamental importance. Exact versions of the schema theorem now take this into account—indeed, they amount to re-writing the evolution equation $p(t+1) = \mathcal{M} \circ \mathcal{F}(p(t))$ by restricting it to subsets of chromosomes. Secondly, and more significantly, the effect of selection on a schema is itself a *dynamic* quantity. This is because the average fitness of schemata (and of the population as a whole), changes from time-step to time-step. Thus, what might qualify as a 'building block' in one generation might not be one in the next generation. The building blocks change all the time. It therefore makes no sense to envisage a genetic algorithm as working with a fixed set of blocks, which it assembles into high quality solutions. Such blocks do not exist, unless they are artefacts of a particular fitness function. Instead, we should realise that, after putting together what are currently perceived to be good schemata, a whole new set of building blocks is created. This pool of building blocks changes *at each time step*. If we want to retain the terminology of the original GA theory, we might therefore refer to this picture of how a GA works as a *dynamic building block hypothesis*.

The weakness of the original, static view of building blocks can be clearly seen in the notion of a 'deceptive' function. This line of research aimed to capture one of the reasons why GAs sometimes fail on fitness functions. The argument was that, since GAs work by putting together good small building blocks to make better ones, then the algorithm would be fooled if in fact the optimum was not made up of these building blocks. A function that had this property was deemed to be deceptive. The term was formally defined in terms of schemata fitness averages taken over the Universe. (Recall from Chapter 3 that a function is called *fully deceptive* if the average fitness of any schema of which it is a member is less than the average fitness of that schema's bitwise complement.) We emphasize again that the averages are here computed for the Universe and not with respect to a particular population, and it is this that is the Achilles heel of the notion of deception. As explained in Chapter 8, it is highly unlikely that the *estimates* of schema fitness from a given population will be reliable indicators of the actual schema fitness at all. Moreover, no account whatsoever is taken of the effects of crossover and mutation, except in the vague belief that they help to assemble building blocks.

We have discussed these points at length earlier, but here is one more example. Suppose we have strings of length 3 and the function:

10.1. THE DEMISE OF THE SCHEMA THEOREM

$$f(0\,0\,0) = 6$$
$$f(0\,0\,1) = 1$$
$$f(0\,1\,0) = 1$$
$$f(0\,1\,1) = 4$$
$$f(1\,0\,0) = 1$$
$$f(1\,0\,1) = 4$$
$$f(1\,1\,0) = 4$$
$$f(1\,1\,1) = 5$$

It is easy to check that the average fitness of the schema $(0\,0\,*)$ is 3.5 whereas the fitness of $(1\,1\,*)$ is 4.5. Similarly the fitness of $(0\,*\,*)$ is less than that of $(1\,*\,*)$. The 'building blocks' will tend to lead to the construction of $(1\,1\,1)$, which is not optimal. The problem is therefore 'deceptive' and the prediction is that a GA will perform badly on it. There are two errors here. Firstly, as we have seen, different parameter settings for a GA (such as mutation rate, population size) can produce very different behaviour on the same fitness function. Chapters 8 and 9 have argued extensively that we simply cannot predict GA behaviour from the fitness function alone. Secondly, the fact that schemata fitnesses are averaged over the Universe means that population dynamics have not been taken into account. Consider the fitness function g defined by

$$g(x) = \begin{cases} 12 & \text{if } x = 0\,0\,0 \\ f(x) & \text{otherwise} \end{cases}$$

Calculating the schema fitnesses for g shows that it is not deceptive at all. But the only thing that has changed is the fitness of the optimal solution. Therefore, a GA running on f and on g will behave identically up to the moment that the optimum is discovered—the fact that one is 'deceptive' and one is not has no effect whatsoever on the behaviour of the algorithm up to that point! The problem, of course, is that deception is defined statically, and GA behaviour is intrinsically dynamic.

Unfortunately, despite what we now know to be the case, appeals to building blocks, and to notions of deception are still made in GA research. The main reason seems to be that these ideas are intuitively very attractive. They are used in the construction of new algorithms and operators into which people build a great deal of their intuitions, garnered from many years' experience with evolutionary systems. There is nothing wrong with using intuitions to try to improve algorithms, as long as it is realised that they require a solid theoretical foundation to make sure that our intuitions are correct.

10.1.1 Schema theorem redux?

What then is the current status of the schema theorem? The last few years has seen something of a revival in interest in schema theorems [187, 188, 272, 273]. If what we have said above is valid, it might appear that this work is misguided—but in fact such an assessment would go too far. What has emerged is a more sophisticated understanding of the place of schemata in our understanding of a GA: schemata exist, but while Holland (and Goldberg following him) characterized them essentially in terms of fitness, it turns out that they have really little to do with fitness.

The schema theorem exists within the exact dynamical systems model as a way of coarse-graining the equations of evolution. The remarkable thing about schemata is their intrinsic relationship to mixing (crossover and mutation), a fact observed by Lee Altenberg [5], and later proven independently by Michael Vose and by Chris Stephens. Schemata are natural subsets on which mixing operators work. The fundamental idea is of compatibility: if we wish to aggregate strings and ask questions about the behaviour of such aggregates, it is important that the behaviour should be the same whether we aggregate first and apply the operators to the aggregates, or apply the operators to strings and aggregate afterwards. This can be shown to be true of Holland's schemata for binary strings: amazingly, it can be further shown [293] that these are the *only* aggregates that are compatible with the mixing operators.

The effects of these operators can thus be projected onto families of competing schemata in an entirely natural way. Unfortunately this cannot be done for selection, since the fitness of a schema keeps changing. This remains one of the fundamental open problems in understanding GA dynamics. Unless it is solved, the schema theorem cannot be used to explain the behaviour of the genetic algorithm.

10.2 Exact Models and Approximations

Strictly speaking, the exact dynamical systems model describes what happens in an exact sense only in the case of an infinite population, and it is tempting to criticise it on those grounds. A practitioner might well remark

10.2. EXACT MODELS AND APPROXIMATIONS

that the global optimum is sure to exist in an infinite population, and notions of 'finding' it are therefore vacuous. Yet although Cantor's concept of the infinite met with much suspicion in its early days, infinity is now frequently invoked within mathematics to help us understand the finite. Cantor's 'paradise' is an essential concept for modern mathematics[1].

The dynamical systems model gives us the exact evolution equations for the genetic algorithm. In the limit as population size increases, these equations do describe the trajectory that a population will follow. For finite populations they tell us the *expected* next population, and by extension, the probability distribution (over all populations) for the next generation. This enables us in principle (and for small problems, in practice [59]) to construct the transition matrix for the Markov chain. Qualitatively, the model still tells us a lot about how we should expect a 'generic' GA (using selection, crossover and mutation) to behave, even in the finite case. We expect the population to move quickly to a metastable state close to a fixed point of the infinite population system, and stay there for a time. After a while the population will jump to a different metastable region. This phenomenon is often observed, and has been described as the occurrence of *punctuated equilibria.*[2]

It is also possible (depending on the fitness function) that there may be whole metastable regions within which the population will wander randomly, before eventually escaping. Such regions are sometimes known as *neutral networks*. The location of these metastable states and neutral networks can be found by examining the dynamics and fixed points of the infinite population equations. However, the length of time a population will stay in such a state, and the probability of its switching to another are dependent on the population size in a way that is not fully understood. This is one of the important open research issues at the moment. Another is the problem of solving for fixed points for a genetic algorithm with crossover. If there is no crossover, the problem reduces to the comparatively simple one of finding the eigensystem of the selection and mutation matrices. Even in this case, the effect of having different population sizes can only be described qualitatively—there is no practical way at present to calculate these effects exactly.

---

[1] It was another great mathematician, David Hilbert, who remarked of the concept of infinity that 'from the paradise created for us by Cantor, no one will drive us out'.

[2] The term was originally popularized by Eldredge and Gould [67] in their first challenge to the gradualist Darwinian consensus.

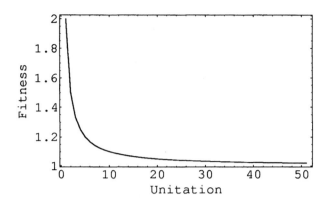

Figure 10.1: Function of unitation $f(x) = 1 + \frac{1}{1+u(x)}$.

10.2.1 Dynamical systems model versus 'building blocks'

We give here an example that nicely contrasts the power of the dynamical systems model with the intuitions of 'building block' theory. Suppose our fitness function is a function of unitation (see Appendix A for more on unitation) on strings of length 50, defined by

$$f(x) = 1 + \frac{1}{1 + u(x)}$$

where $u(x)$ is the number of ones in a string. This function is shown in figure 10.1. It is a unimodal function in the Hamming landscape, and it is simple for a hill-climbing algorithm to solve it.

We shall use a GA with proportional selection and mutation (with a rate $\mu = 0.02$). A naive interpretation of the schema theorem would suggest that we should check whether or not schemata to which the optimum string belongs are fitter than their competitors. Since the more zeros a string contains, the fitter it is, this is definitely the case. In other words, the function is not at all 'deceptive'. The prediction would be, then, that the building blocks containing zeros would be assembled and the optimum found. Now we shall use the dynamical systems approach to work out what really happens.

Constructing the mutation and selection matrices, we calculate the leading eigenvector. This gives us the infinite population fixed point. It is shown in figure 10.2. It is clear that the population distribution at the fixed point is approximately binomial—i.e., what we could expect from a population of strings chosen uniformly at random. The corresponding eigenvalue tells us

10.2. EXACT MODELS AND APPROXIMATIONS

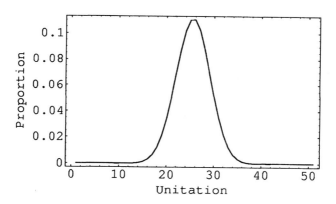

Figure 10.2: Fixed point for fitness function $f(x) = 1 + \frac{1}{1+u(x)}$ with proportional selection and mutation, $\ell = 50, \mu = 0.02$. The vertical axis records the proportion of the population with different values of $u(x)$.

that the average fitness of this population is 1.04.

We can run an actual GA on this problem with population 200. To make things more interesting we shall start with a population that contains exactly 200 copies of the optimal string. The average fitness of this population is, of course, 2. As the system evolves, we plot the average fitness at each generation. The results of a typical run are shown in figure 10.3. It is clear that the population is headed towards the fixed point, and we see that the average fitness *decreases* dramatically from time-step to time-step. We start with the optimal solution (and only the optimal solution) and finish with a random population. Evolution appears to be going 'backwards'! This result completely goes against the naive interpretation of the schema theorem, but is exactly predicted by the dynamical systems model. (Exercise for the reader: what happens when crossover is added?).

10.2.2 Finite population models

There are some things that can be said in general for finite population GAs by appealing to general Markov chain theory. It is known that an élitist selection strategy will allow the GA to converge to the optimal solution. However, it is not known how long this will take. There are also convergence theorems for various annealing schedules for mutation and selection strength. For the simple GA (that is, with proportional selection and no élitism) the population will not converge: it will visit every possible state infinitely often. There is, however, a limiting distribution, which tells us how

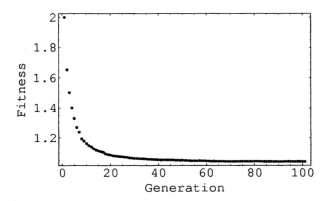

Figure 10.3: Average fitness as the genetic algorithm evolves. Starting with the optimal population, the GA actually evolves 'backwards' to a random population.

frequently each population will be visited. Again, calculating this distribution is difficult, owing to the huge number of possible populations, even for small problems. What we really want is some mathematical way of describing the limiting distribution in terms of the underlying infinite population dynamics and the population size. The finite population can be viewed as sampling the dynamics in a stochastic fashion, and deviations from the expected behaviour can be thought of as arising from sampling errors.

There are a couple of hopeful lines of investigation here. One is to use a coarse-graining of the variables to reduce the number of degrees of freedom of the system. This might be done by aggregating states (as in [268]) or through the use of schemata, as described above. A combination of these methods might prove fruitful. The second idea is to pursue the statistical mechanics approach. Here we describe the population as an infinite series of statistical measures (mean, variance, etc.). Truncating this series will provide what we hope is an accurate approximation in (preferably) few variables. While there has been some success in applying this approach to certain fitness functions, it is undoubtedly a difficult task in general, since there is no systematic way of producing these approximations, nor any theory to describe the errors that this technique introduces. Formally linking the dynamical systems model with the statistical mechanics approach would be a big step forward.

10.3 Landscapes, Operators and Free Lunches

It should be clear by now that one of the most important and difficult problems in understanding genetic algorithms is to understand the interaction of the mixing operators with the structure of the fitness function. Attempts to characterise fitness functions on their own, usually by means of a Walsh (or more generally, a Fourier) decomposition, have led to some very interesting ideas on measuring and predicting *epistasis*, but it is clear (as described in Chapter 8) that omitting explicit reference to the search algorithm and its operators is to ignore the fundamental determinant of the difficulty of a problem instance. The 'simple' *Onemax* function, for example, can be made very difficult for a hill-climber by changing the search operator from bit-flipping to the 'crossover-like' CX.

This interaction can be interpreted in general by means of the concept of a *landscape*, as described in Chapter 9. While a lot of superb research has been done on trying to understand fitness landscapes—in particular, Peter Stadler's seminal work [269], which has revealed the centrality of the Walsh/Fourier basis—it is not obvious how this information helps with predicting GA behaviour. The spectral decomposition of a landscape can tell us *a posteriori* why one operator works better than another, but finding accurate *estimates* of the Walsh coefficients in order to obtain a satisfactory *a priori* measure of performance is very hard, even if we make restrictive assumptions about the nature of the decomposition.

In fact the No Free Lunch theorem tells us (among other things, as discussed in Chapter 4) that, in general, the structure of a fitness landscape cannot be inferred from a sample of points unless some additional knowledge about the function is given. A simple version of the NFL theorem illustrates this point. Suppose we have a set of playing cards face down on the table. The task is to turn over an ace as soon as possible (any ace being an 'optimum'). The rules allow us to turn over one card at a time. At a certain stage of the game, several cards are face up, but none of them is an ace. This certainly gives us some information about where the aces are not to be found! But does it tell us anything that will help us find an ace more quickly? Obviously not. We have roughly the same situation when a GA (or indeed any search algorithm) is trying to optimize a discrete fitness function. At any stage of a run, a certain number of points have been examined. None of them is the optimum[3]. Can we use the information gathered from

---

[3] Actually, the real situation may be even worse than in this illustration. We do at least recognize an ace when we see one, but in optimization, we often don't even know what the optimum looks like, so we can't be sure that the optimum is still to be found.

these points to tell us where we should look next? No! The optimum could be anywhere. Only if we have some prior knowledge of the fitness function can we use the information we have to predict something about the unseen points. For example, in the card game we might be told that aces tend to be placed near picture cards. If we have seen some picture cards then this might guide us as to where to look next.

It is sometimes argued that the NFL theorem does not apply to 'real life' since most fitness functions are completely random, and the ones we normally encounter are not. More precisely, most fitness functions are not compressible: they cannot be implemented by short computer programs, only by large look-up tables. (This is the essence of Tom English's discussion of NFL from the perspective of information theory, as described in Chapter 4.) However, the fitness functions we usually meet tend to be implemented as short computer programs. We are therefore only dealing with a small subset of all possible functions. The NFL theorem does not apply, then, as long as we can figure out some way of making use of this prior information. That this is a possibility has been demonstrated by Droste *et al.* in their 'free appetizer' paper [63]. However, there is as yet no practical way of making use of this information, nor of characterising structural properties of compressible fitness functions in terms of landscape decomposition.

The argument of Igel and Toussaint [132] goes a little further, by suggesting that the sort of properties that a 'real' fitness landscape might possess (such as *not* having the maximum possible number of optima) characterise precisely those classes of functions where NFL does not hold. As discussed in Chapter 8, this amounts to an assumption about the world—an 'article of faith', that the real world is not totally arbitrary or capricious, but that it has sufficient structure and order to be 'interesting', but not so much as to be completely incomprehensible[4].

10.4 The Impact of Theory on Practice

One of the main ways in which theory can help practitioners is in challenging their assumptions and intuitions about what it is they are doing. Evolutionary algorithms are highly complex. This means that our intuitions often turn out to be wrong or at least misleading. For example, not so long ago genetic

---

[4]As remarked in [223], this assumption is nothing new in science. Hooykaas [129] makes an interesting case that pre-Reformation world views, in effectively deifying Nature, made the development of modern science almost impossible. It had to await the Protestant (and especially the Calvinist) faith in a non-capricious God whose world worked according to providential 'law'.

10.4. THE IMPACT OF THEORY ON PRACTICE

algorithms were being touted as general purpose black box optimisers that were guaranteed to perform better, on average, than random search. Indeed, it was said that the way GAs searched a space was in some sense optimal! We now know, of course, that this is wrong: no search algorithm is better than any other on average. However, there may be some class of problems for which they are particularly suitable, but characterising that class is an open problem. We may believe that they *do* perform better than random search for 'real-world' problems, but we cannot make such beliefs rigorous unless we can characterise both GAs and problem classes better than we can do at the moment.

Practitioners would like us to come up with some simple rules of thumb for setting parameters, such as population size. But we now know that the relationship between, for example, population size and algorithm behaviour is, in general, very complex. There will not be any magic formulae for setting parameters that will enable us infallibly to solve difficult problems: some problems are just difficult! Again, though, it is possible that something may be done for certain restricted classes of problem. Some progress has been made on this, but only for toy problems so far.

Some specific things can be said. Here are a few examples:

1. For a guarantee that an algorithm will eventually converge (in some sense) to the optimum, then some form of élitist strategy or steady-state algorithm should be used. Alternatively, the mutation rate can be reduced slowly, while increasing the strength of selection (according to a schedule that is known).

2. In order to optimize a function of real-value parameters using a GA, then it is typically better to use a Gray code representation rather than a straightforward binary encoding, unless the fitness function is pathological. (Then again, perhaps an Evolution Strategy approach should be taken at the outset!)

3. Increasing the population size will cause the GA to follow the infinite population dynamics more closely, which means that the population will spend most of its time in the vicinity of fixed points of those dynamical equations. However, this may or may not be a good thing, as far as search is concerned.

4. Increasing the mutation rate pushes the fixed points further into the simplex, so that populations will tend to be a mixture of a large number of individuals. Again, this may or may not be a good thing.

5. The role of crossover is still not well understood. We know that crossover can help the population to jump over certain kinds of gaps in the search space. We have examples of problems where recombination gives significant speed-up. Designing good recombination operators is problematic, however, unless we happen to know which properties ought to be preserved during the search process. In this case, there are results that may help, but they rely on having a good understanding of the nature of the search space.

6. It is important also to bear in mind the dual effect of crossover: recombination is a general way of combining common elements of two (or more) parents, but it also involves the application of a specific search operator. We need to be sure that these two effects are working in harmony.

7. In the case of variable-sized structures (for example, in genetic programming) then the choice of crossover and mutation operator can considerably bias the search process towards certain sizes. Apparently small changes in the representation of the problem can have similarly dramatic effects (for example, changing the number of possible terminals in a GP problem).

8. In general, the more knowledge of the problem that we can build into the algorithm, the better. In particular, we should try to make the representation and operators as 'natural' as possible for the problem.

One of the tasks of the theorist is to critique models of GA behaviour, in order to find out their validity and scope of application. We should be especially careful about models of GAs that are designed around particular classes of problems. These are often constructed by researchers trying to capture certain intuitions about GA behaviour. We might describe them as 'engineering' models, whose main aim is to describe the intuition in such a way that a more efficient algorithm can be written. There is nothing wrong with this in itself. However, we have already seen how badly wrong our intuitions can be when it comes to evolutionary algorithms. If a theory is built around one class of problems, then the question must be asked as to the extent to which the model generalizes to other classes. For example, a large quantity of research has concentrated almost exclusively on concatenated trap functions. A typical trap function on three bits is:

$$f(0\,0\,0) = 4$$

10.4. THE IMPACT OF THEORY ON PRACTICE

$$f(0\,0\,1) = 1$$
$$f(0\,1\,0) = 1$$
$$f(0\,1\,1) = 2$$
$$f(1\,0\,0) = 1$$
$$f(1\,0\,1) = 2$$
$$f(1\,1\,0) = 2$$
$$f(1\,1\,1) = 3$$

This looks rather like the *Onemax* function, but with a global optimum inserted at string (0 0 0). (More details on such problems can be found in Appendix A.) The idea is then to concatenate a number of these together to make a problem on a larger set of bits, with many sub-optima. Such a function seems to have a number of natural 'building blocks', namely the solutions to the individual trap functions. Of course these are purely properties of the way these particular functions are defined. The theory that studies these functions tries to generate rules for parameter settings such as mutation rate, population size and so on. While the formulae that are derived may capture some intuitions about how a GA will behave on these functions, they are given in terms of such quantities as 'building block size', a term which is (naturally!) well-defined for the family of functions under consideration. Unfortunately, it is not well-defined for any other class of functions. This means that the results cannot be assumed to apply outside the classes of concatenated-trap functions. We must be very careful, then, as to how much one reads into the intuitions underlying the models.

Theorists are also concerned with the quality of research produced by practitioners. Practical experimental results should help to suggest new hypotheses to test and new avenues to explore from a theoretical angle, but this is not often the case. The number of papers devoted to GAs is enormous—and increasing—but sadly, too many of them have titles like 'A New Genetic Algorithm for the Multi-Resource Travelling Gravedigger Problem with Variable Coffin Size', describing a minor variation on the GA theme that beats the competition by a rather insignificant amount on a few instances of a highly specific problem. If a reader wishes to invent a new evolutionary algorithm for a particular problem class, he or she might like to consider the following guidelines:

1. Results should be compared with other standard algorithms over a good variety of instances of the problem. (As a minimum, we would hope that the GA at least beats random search—something that should

be fairly easy to check.)

2. Observed differences in performance should be checked for statistical significance.

3. There should be an awareness of the theoretical limitations of finding approximate solutions to the problem.

4. The sensitivity of results to changes in parameter settings should be tested (preferably by following a principled statistical experimental design).

5. The nature of the problem class should be characterised so that what it is that makes the algorithm succeed can at least be conjectured.

6. The types of problems on which the algorithm will fail should also be investigated.

The last two points are particularly important. We know that an algorithm cannot be good for every fitness function (because of the No Free Lunch theorem). It is therefore inadequate to design a new variant evolutionary algorithm and show that it beats some other algorithm on a small set of benchmark problem instances. We need to know why it manages to beat the other algorithm. We would also like to know what its weaknesses are—it must have some! Only when we are truly aware of an algorithm's strengths and weaknesses can we begin to say that we understand it.

10.5 Future research directions

Now that a formal mathematical description of genetic algorithms is in place, and we are building on a firm foundation, there are many important and exciting research questions that need to be addressed. Some of these are rather technical, and would require a detailed knowledge of a particular approach. Others are wide open to anyone who can come up with a good enough set of ideas. We list a few of these areas which we think are significant.

10.5.1 Technical issues

There are several important technical issues in the dynamical systems theory that need further study. Some of them may seem fairly obscure, but there are some which relate to obvious gaps in our understanding of GAs. One such issue is the case of a GA with proportional selection and crossover

10.5. FUTURE RESEARCH DIRECTIONS

but no mutation. It is clear that the vertices of the simplex (that is, populations containing copies of just one item) are fixed points. In particular, a population containing only the optimum solution is an *asymptotically stable* fixed point [291]. This means that if we perturb the population slightly, the GA will converge to the fixed point again. For binary strings, Hamming landscape *local* optima are also asymptotically stable fixed points (as discussed in Chapter 9). What is not clear is if there are any asymptotically stable fixed points in the interior of the simplex. That is, we would like to establish if it is possible (or not) to construct a population that, for a given fitness function, has the following properties:

1. It is a fixed point: applying selection and crossover takes us back to the same population.

2. It is inside the simplex: it contains a copy of every element of the search space, in various proportions.

3. It is asymptotically stable: on changing the composition of the population slightly and running the GA, it will converge to the fixed point once more.

It is known that if the fitness function is linear, there cannot be such a population, but the status of the non-linear case is unknown. This issue has been posed in terms of the infinite population model, but it is has relevance for the finite population case. The question is simply: what happens when we run a GA without mutation? The conjecture is, roughly, that we typically always see the population converging to a uniform population, and that such states are absorbing for finite populations.

Another open issue is to try to characterise those types of mutation and crossover which guarantee that the infinite population GA always converges to a limit. It is known that there are some mutation schemes (admittedly, fairly weird ones) for which the dynamics of the population becomes periodic, or even chaotic. We would like a theorem that says that as long as mutation and crossover are fairly 'sensible' (where this term needs to be characterised formally), then the GA will always converge to a limit point. This question also has practical relevance. If the infinite population dynamics are periodic or chaotic, then we should expect the finite population dynamics to be pretty strange too.

10.5.2 Finite populations

The dynamical systems model provides exact equations for the evolution of GAs in the infinite population limit. And because it also provides us with the distribution over all possible populations for the next generation of a finite population, we can use the equations to calculate the transition matrix. However, working out the stationary distribution of the corresponding Markov chain is impracticable for any reasonably sized system.

A major open question, then, is to how to relate the stationary distribution of a finite population to the infinite population dynamics. At the moment we use a qualitative argument, that the finite population will tend to spend most of its time in regions where the force of $\mathcal{G}$ is small. We can find some of these regions if we can analyse the fixed points of $\mathcal{G}$. However, the amount of time spent in these metastable states and neutral networks, and the probabilities of the transitions to other such states, cannot be calculated using current methods. There are some technical theorems which describe the way in which such probabilities converge to the infinite population dynamics. This means that something can be said about the behaviour of populations that are sufficiently large. However, for small populations, the problem is wide open.

The question of *where* the population is likely to go is also of fundamental importance in the context of optimization. (And as we have been saying since Chapter 1, optimization is *the* context in which most practitioners are interested, although we are well aware that it is not the only one.) While the dynamical systems model can tell us in general, qualitative, terms how a GA behaves, practitioners are probably more interested in the chances of finding the global optimum (or even a 'good' local optimum) and in how long it will take to get there. These questions can be answered in principle by a Markov chain approach, but in practice it is usually impossible.

10.5.3 Systematic approximations

One source of progress on the question of finite population behaviour comes from the statistical mechanics approach. For the systems that have been analysed, it is possible to add finite population effects in terms of sampling errors. However, it is not clear how this approach can be adapted to more general problems. In fact, it would be a major step forward if one could find a systematic way of deriving approximate models such as the statistical mechanics ones from the exact, infinite population equations.

We could start by asking what, in general, is the average fitness of a population after applying the operator $\mathcal{G}$? Suppose that we have proportional selection and mutation, but no crossover. We know that

$$p(t+1) = \frac{USp(t)}{\kappa_1(p(t))}$$

where U is the mutation matrix, $S = \text{diag}(f)$ and κ_1 is the mean fitness. Define the linear functional $h: \mathbb{R}^n \to \mathbb{R}$ by

$$h(x) = \sum_i x_i,$$

so that we have $\kappa_1(p) = h(Sp)$. Therefore

$$\kappa_1(p(t+1)) = \frac{h(SUSp(t))}{\kappa_1(p(t))}$$

Now if we suppose that mutation is small, we can write the mutation matrix as $U = I + M$ where M has very small entries. Substituting gives us

$$\kappa_1(p(t+1)) = \frac{h(S^2 p(t)) + h(SMSp(t))}{\kappa_1(p(t))}$$

It can be checked that the variance of the fitness satisfies $\kappa_2(p) = h(S^2 p) - [h(Sp)]^2$, and so after some rearranging

$$\kappa_1(p(t+1)) = \kappa_1(p(t)) + \frac{\kappa_2(p(t))}{\kappa_1(p(t))} + \frac{h(SMSp(t))}{\kappa_1(p(t))}$$

We see, then, that the mean fitness increases by an amount equal to the variance divided by the mean, plus a correction term, which should have relatively small values (since M is small). To continue to develop the statistical mechanics model along these lines we would need to

1. develop equations for the higher cumulants;

2. put a bound on the error terms;

3. work out a scheme for including crossover;

4. add corrections for finite populations.

If this could be done, we would then have a generic scheme for deriving approximate equations in a small number of variables.

10.5.4 Landscapes and operators

One of the great 'missing links' in GA theory is the connection between the effects of selection and mixing (crossover and mutation). Mixing is understood fairly well on its own, but the interactions that take place with selection are difficult to analyse. We have seen that a considerable amount of effort has gone into understanding the structure of fitness landscapes, in the hope that this will tell us something about the effect of operators on a problem. A particularly intriguing line of research would be to examine a fitness landscape using a discrete Fourier analysis. In the binary string case, this corresponds to taking the Walsh Transform of the search space, which reveals the interdependencies of the different bits in composing the fitness function. However, as we have seen in Chapter 6, the Walsh transform is also of fundamental importance in understanding mixing in a GA. The hope is, then, that the two lines of research might join up and the full GA (selection plus mixing) be understood more thoroughly.

The problem at the moment seems to lie in the following. Even if we have a Walsh transform of a fitness function, this does not simplify the action of selection as an operator. Consider, for example, the fitness function on two bits:

$$f(0\,0) = 1$$
$$f(0\,1) = 2$$
$$f(1\,0) = 2$$
$$f(1\,1) = 3$$

The Walsh transform for this case is the matrix

$$\boldsymbol{W} = \frac{1}{2}\begin{bmatrix} 1 & 1 & 1 & 1 \\ 1 & -1 & 1 & -1 \\ 1 & 1 & -1 & -1 \\ 1 & -1 & -1 & 1 \end{bmatrix}$$

This gives a Walsh transform[5] of the fitness landscape of

$$\boldsymbol{Wf} = (4, -1, -1, 0)$$

which indicates that there are no interactions between the bits (the last entry is zero). However, if we now look at the selection matrix in the Walsh

---

[5] As remarked in Chapter 6, these values are a multiple of 2 times the Walsh coefficients as defined in Chapter 3.

basis we have

$$W \operatorname{diag}(f) W = \frac{1}{\sqrt{2}} \begin{bmatrix} 4 & -1 & -1 & 0 \\ -1 & 4 & 0 & -1 \\ -1 & 0 & 4 & -1 \\ 0 & -1 & -1 & 4 \end{bmatrix}$$

and so we have turned a diagonal matrix into a much more complicated one. One possible way forward would be if there was sufficient symmetry within the fitness function that the number of variables needed to describe it could be reduced to a minimum. This might at least provide a way of building a tractable set of evolution equations.

10.5.5 Avoiding No Free Lunch

There are two basic steps required to avoid falling foul of the No Free Lunch theorem. Firstly, we need to characterise a subset of functions that we intend our algorithms to optimize. Secondly, knowledge derived from this characterisation needs to be built into the algorithm so that this structure can be exploited. By adding domain knowledge to an algorithm we are essentially 'paying for lunch'.

We have seen that one approach to characterising functions is in terms of their compressibility. A popular method for trying to accomplish this is to look at the Walsh transform of the fitness function. If all the high-order interactions are zero (or near zero) then the function has low epistasis and might be amenable to easy optimization. Unfortunately, as we saw in Chapter 8, this is not a very good criterion. If only the order one Walsh coefficients are non-zero, then the problem is linear and easy to solve (and we would not need a GA anyway). However, if the order two coefficients are non-zero then we are, barring a uniformly 'benign' form of epistasis (as discussed in Chapter 8), in the class of NP-hard functions. It is possible that for a given fitness function, some other basis may be found in which it has a simpler (near linear) structure, but the problem of finding such a basis remains.

The most important lesson of No Free Lunch is that we have to put some domain knowledge into the algorithm in order to gain some leverage. There has been some interesting work done on the design of particular crossover operators for certain combinatorial problems [87]. A systematic method for designing appropriate mixing operators would be extremely helpful.

10.5.6 Computational complexity

An important aspect of understanding any algorithm is to analyse its running time complexity. For a search algorithm this means either the average time until the optimum is found or the time taken to guarantee that we have a solution that is a good approximation to the optimum. Because of the complex nature of genetic algorithms it is very hard to prove such results in any generic fashion. Ralf Salomon [251] has discussed the time complexity of GAs for separable functions, where we could in principle use a deterministic algorithm that solves for each gene independently. His analysis suggests that there is a $\mathcal{O}(\log \ell)$ penalty to pay in terms of computational effort (e.g., if a deterministic algorithm was $\mathcal{O}(\ell)$, a GA for the same function would be $\mathcal{O}(\ell \log \ell)$), and that this can be attributed to the stochastic nature of an evolutionary algorithm.

However, there has also been some recent success in designing particular fitness functions to demonstrate the utility of certain features of the GA. Thomas Jansen and Ingo Wegener [133] have constructed fitness functions for which, provably:

- a GA with crossover outperforms a GA with no crossover, in the sense that the crossover algorithm finds the optimum in polynomial time (as opposed to exponential time required by the selection and mutation only algorithm);

- a GA with a population outperforms a single-individual evolutionary algorithm (in the same sense).

Related work by Xin Yao and Jun He has produced a set of criteria against which evolutionary algorithms can be tested to guarantee polynomial running time [113][6]. An interesting weakness of this work is that it has to assume that the effects of crossover are not too deleterious. Trying to capture the *advantages* that crossover might offer is difficult.

In terms of a stochastic process, we are interested in the expected *first hitting time* of the Markov chain. That is, we wish to know how long it will take, on average, before we see the optimum. Part of this calculation may involve looking at *order statistics*. These are the probability distributions associated with the best, second best, and in general the kth best element in the population. Some analysis along these lines was done by Günter

---

[6]Ingo Wegener [personal communication] has recently pointed out to us that there is an error in one of the theorems in [113]. However, this does not detract from the general approach.

Rudolph [247] but calculations for specific cases quickly become intractable. The development of generic techniques for estimating the running time of GAs is a very important research topic, and one which will help establish those classes of functions for which GAs will perform efficiently.

10.6 The Design of Efficient Algorithms

Before we can be in a position to design efficient genetic algorithms tailored to specific classes of fitness function, we shall have to understand much more about the dynamics of populations, and about their relationship to the structure of the fitness landscape. A rigorous mathematical theory is beginning to emerge in which our ideas and intuitions can be formally expressed and either proved or refuted, and we have tried to cover this ground as comprehensively as possible.

However, there is still much to be done. Fortunately, the complexities of the GA make it a most fascinating subject of research. The development of a theory of GA behaviour would be worthwhile even if these algorithms were not to be used in real-world applications. That they are being used in this way is an indication of their great potential. It also indicates the great need to press ahead and develop a more complete theoretical framework, which will be able to guide practitioners, formalise their ideas, and challenge their assumptions.

Without doubt, the development of this theory is difficult. Trying to communicate the results and significance of our theoretical advances to the wider research community is challenging. But such advances must be made and effectively communicated, if genetic algorithms are to be used effectively to their full potential. It is our hope that this book will have aided that communication, and will have stimulated researchers with a good mathematical background to become involved.

Appendix A

Test Problems

One way in which GA researchers have tried to understand how the algorithm operates is to investigate the performance of the algorithm on some common benchmarks or test problems. This practice was initiated by Ken De Jong in his thesis [55], but his concern there was mostly in how well a binary GA could do in solving real-valued problems. However, if our primary interest is in how a GA works, introducing a genotype-phenotype mapping into the problem just complicates matters further, so most of the test problems that have been devised for theoretical investigations are assumed to be a function of binary strings only. As some of these recur frequently, it is sensible to give a brief outline of some of them. Of course, this makes no pretence of being an exhaustive list of all the classes of problems that have been investigated.

A.1 Unitation

Many test problems are based on the idea of 'unitation'. The unitation of the vector $x \in \{0,1\}^\ell$ is

$$u(x) = \sum_{i=1}^{\ell} x_i.$$

The advantage of unitation functions is that all solutions belong to one of a set of $\ell + 1$ equivalence classes $C_0, \ldots, C_\ell$ based on the number of bits that take the value 1, i.e.,

$$x \in C_k \iff u(x) = k.$$

This effectively reduces the dimensionality of the solution space and thus facilitates analysis. Several examples that illustrate this aspect can be seen in chapters 6 and 7.

Perhaps the simplest function of unitation is the *Onemax* function.

A.2 *Onemax*

The *Onemax* function is also called 'onesmax' and 'ones-counting' by some authors. For a vector $x \in \{0,1\}^\ell$, *Onemax* is

$$f(x) = u(x) = \sum_{i=1}^{\ell} x_i.$$

This is a very easy linear function of the variables x_i, very quickly optimized by a bit-flipping neighbourhood search approach. Although it is claimed by many to be 'GA-easy', it turns out to be harder than might be expected, for reasons that have been explained in chapter 9.

A.3 Trap functions

Trap functions also use the idea of unitation. They were designed by Deb and Goldberg [54] to pose greater problems than *Onemax* can. A local minimum class is selected at $u(x) = z$, and two slope parameters a and b define the function:

$$f(x) = \begin{cases} a\, \frac{z - u(x)}{z} & u(x) \leq z; \\ b\, \frac{u(x) - z}{\ell - z} & \text{otherwise.} \end{cases}$$

This has local maxima at $u(x) = 0$ and $u(x) = \ell$. The relative magnitudes of these maxima can be tuned by choosing a and b appropriately.

A.4 Royal Road

The *Royal Road* functions form a class devised by Mitchell et al.[169] in an attempt to demonstrate the BBH. The idea is to construct a function from disjoint schemata, so that the conditions for the BBH are fulfilled. Several variants have been proposed; the recursive algorithm in Figure A.1 is fairly generic.

The idea is that the function's value depends on the number of *complete* basic building blocks (BBs) it gets right. The exact form of the functions

A.4. ROYAL ROAD

Defining the building blocks:

Assume that $\ell = 2^m$ for some integer m.
For a binary string x of length ℓ,
for $i = 1, \ldots, 2^m$ **do**
    Set $B_i^{(0)} = x_i$
endfor
for $k = 1, \ldots, m-1$ **do**
    for $i = 1, \ldots, 2^{m-k}$ **do**
        Define $B_i^{(k+1)} = B_{2i-1}^{(k)} \mid B_{2i}^{(k)}$
        where the symbol $\mid$ means *concatenation*.
    endfor
endfor

Computing the function:

$$f(x) = \sum_{k=1}^{m} \sum_{i=1}^{2^{m-k}} f_k(B_i^{(k)})$$

Figure A.1: Generic definition of Royal Road functions. While such functions could be devised for values of ℓ that are not powers of 2, such an assumption simplifies matters greatly. The functions f_k are used to encourage different types of behaviour.

$f_k(\cdot)$ in Figure A.1 depend on the type of Royal Road that is required. In the simplest case (this is the 'R1' function of [169]), f_1 is the function

$$f_1(B) = \begin{cases} c_1 & \text{if } u(B) = |B| \\ 0 & \text{otherwise} \end{cases}$$

and $f_k(\cdot) = 0$ for all other values of k. In other words, the 'optimal' BB of any size is assumed to be a block of 1s, and the discovery of blocks of 1s of ever-increasing size produces a staircase that leads inevitably to the global optimum.

More complicated functions can be devised from the same scheme: the 'R2' function proposed that constant 'bonuses' $\{c_k\}$ could be given to *contiguous* blocks in order to encourage their formation; e.g., the contribution of $(1, 1, 1, 1)$ to the fitness would be just $2c_1$ in the case of R1, but it would be $2c_1 + c_2$ for R2.

A.5 Hierarchical-IFF (HIFF)

The trouble with Royal Roads as a test problem is that they are too easy for hill-climbers, being easily separable; in fact much easier than for a GA. To pose a more severe test, Watson *et al.* [298] suggested a variation. Their 'Hierarchical-IFF' (HIFF) functions are constructed in an identical fashion, but the assignment of partial fitnesses is very different. For HIFF functions, blocks of 0s and 1s alike get credit. However, when 'optimal' blocks of 0s and 1s are combined at the next higher level, they get nothing: only homogeneous blocks will do.

A.6 Deceptive functions

David Goldberg [96, 97] developed the idea of deception from a schema processing perspective, as explained in chapter 3. Defining a deceptive function in his terms entails setting up inequalities involving schema averages, and finding values for the Walsh coefficients that are implied by these inequalities. However, there is a simpler approach, as proposed by Darrell Whitley [302].

Goldberg's function has the property that a value of the function at a point with k 1s is always greater than a value at a point with $(k + 1)$ 1s. This suggests the requirement that

$$\min_{x:u(x)=k} f(x) > \max_{x:u(x)=k+1} f(x)$$

Thus in order to generate a deceptive function, $2^\ell - 1$ distinct values for f can be chosen and ranked in decreasing order, and then assigned in rank order to points grouped by *increasing* unitation from $u(\boldsymbol{x}) = 0$ to $u(\boldsymbol{x}) = \ell - 1$. Finally the point **1** is assigned a value higher than that of **0**.

Many applications of these functions only use small values of ℓ. Functions defined over a larger search space are then obtained by concatenating many smaller subfunctions.

A.7 Long paths

The simplest long path described in [130] is the ℓ-dimensional *Root2path* in the landscape $(\{0,1\}^\ell, f, d_H)$, whose construction is as follows. The one-dimensional path is given by the sequence (0 1) where **0** is the starting point and **1** the end point of the path. To construct the $(\ell + 2)$-dimensional path, take two copies of the ℓ-dimensional path. The bits **0 0** are affixed to all elements of the first copy, and the bits **1 1** affixed to those in the second, forming two subpaths. (Thus the Hamming distance between the corresponding points on the subpaths is two.) The order of the elements in the second subpath is reversed and, finally, the end points of both subpaths are connected through a *bridge point*, which begins with either (**1 0**) or (**0 1**). For example, the 3-dimensional *Root2path* is given by the sequence

$$0\,0\,0 \quad 0\,0\,1 \quad 1\,0\,1 \quad 1\,1\,1 \quad 1\,1\,0.$$

It is clear from this construction that the length of the path more than doubles at every recursion. More precisely, if n_ℓ is the length of the ℓ-dimensional *Root2path*, it obeys the recurrence

$$n_\ell = 2n_{\ell-2} + 1$$

whose solution is
$$n_\ell = 3\left(2^{(\ell-1)/2}\right) - 1.$$

If R denotes the set of all binary strings in $\{0,1\}^\ell$ that belong to the *Root2path*, the function $h : R \to \mathbb{Z}$, which returns the position of each element in the path, defines a strictly monotonic increasing function on the path and will serve as an objective function for those points on the path. However, although the length of the path increases exponentially in ℓ it does so more slowly than the cardinality of $\mathcal{X}$. Thus, as pointed out in [130], the *Root2path* becomes a 'needle-in-a-haystack' problem for higher values of ℓ, and it is unlikely that any hill-climber will actually start on the path at all.

To direct the search towards the path, a *slope function* is used. A simple idea is to use unitation again: many functions would do, but the *Zeromax* function (the converse of *Onemax*) is an obvious choice. Since the string **0** is on the path, a hill-climber will be guided towards the start of the path from any initial point. The resulting objective function of the *Extended Root2path* is given by

$$f(x) = \begin{cases} \ell + h(x) & \text{if } x \in R \\ \ell - u(x) & \text{otherwise} \end{cases}$$

Yet longer paths could be devised by a similar process, simply by affixing more bits to the sub-paths at each recursion.

A.8 NK landscape functions

Figure A.2 describes Kauffman's concept of an NK landscape function. As remarked in chapter 9, the default distance metric assumed when using these functions is the Hamming distance, so that the function itself is often referred to as a NK-landscape.

It should be noted that there are two basic versions of the NK-landscape functions; the first is the *random model*, as in Figure A.2, where the bits in set I_i are chosen randomly. The second is the *adjacent model*, where I_i are chosen to be those bits that are in some sense adjacent to i. This still leaves room for some ambiguity: for example, if $K = 4$, the set I_5 might be $\{6, 7, 8, 9\}$. However, if we wanted them to be *nearest* neighbours, we could choose $I_5 = \{3, 4, 6, 7\}$.

A.9 ℓ, θ functions

NK-functions have some interesting and possibly unusual properties [115]; for instance, the number of non-zero Walsh coefficients of a given order r is bounded above by the quantity

$$\min\left\{\binom{N}{r}, N\binom{K+1}{r}\right\} \quad \text{for } r \leq K+1,$$

which implies that for large N relatively few interaction coefficients contribute to the overall fitness. The distribution of the non-zero coefficient values has also been investigated empirically [228], and the evidence suggests that this distribution weights coefficients of different orders in a way

A.9. ℓ, θ FUNCTIONS

Setting up the tables:

For a binary string of length N,
for $i = 1, \ldots, N$ **do**
    Choose K distinct indices $I_i = \{k : k \in \{1, \ldots, N\} \setminus \{i\}\}$;
    Choose 2^{K+1} uniform random variates $u_j, j = 0, \ldots, 2^{K+1} - 1$;
    Store them in table T_i;
endfor

Evaluating the fitness:

Given a binary string S of length N, set $v = 0$;
for $i = 1, \ldots, N$ **do**
    Determine the alleles of S corresponding to the bits in I_i;
    Denoting these values by $z_1, \ldots, z_K$, decode the concatenated bits
    $(i, z_1, \ldots, z_K)$ to an integer r;
    Find u_r, the rth table entry in T_i;
    Set $v = v + u_r$;
endfor
Set fitness $= v/N$.

As an example of the procedure, suppose we have $N = 3$ and $K = 1$. The index sets I_1, I_2, I_3 are $\{2\}, \{1\}, \{1\}$ and the tables T_1, T_2, T_3 have entries $(0.2, 0.4, 0.5, 0.1), (0.6, 0.3, 0.4, 0.9), (0.3, 0.5, 0.1, 0.7)$ respectively. Then the fitness of string (1 0 1) would be found from the entries for (1 0) $\equiv 2$, (0 1) $\equiv 1$, (1 1) $\equiv 3$ from T_1, T_2, T_3 respectively. Thus the fitness of the string (1 0 1) is $(0.5 + 0.3 + 0.7)/3 = 0.5$.

Figure A.2: Description of an NK-landscape function. Note that in this book we have normally used ℓ for the length of a string, but we use N here for consistency with Kauffman's terminology. Note also that this really only defines a function; the term NK-landscape assumes (by default) the Hamming distance metric.

> 1. Define the mean and all main effects by drawing a random variate from some distribution $F(\theta)$, where θ is a parameter vector. Without loss of generality these values will be related to the case where the allele value is 1.
>
> 2. Define k-th order interactions by drawing a random variate from a distribution $F(\theta, k)$—again these are related to the case where allele values are 1s.
>
> 3. Allocate $+$ and $-$ signs to the interaction effects according to some prescription.
>
> 4. (Optional) Finally adjust the 'baseline' by modifying the mean μ so that the string with the minimal value in the Universe has value zero.

Figure A.3: A template for the (ℓ, θ) class of tunable landscapes

that is unlikely to be observed in 'real' functions. To this extent, we should perhaps be cautious in interpreting the behaviour of NK-landscape functions as of more general relevance. The ℓ, θ functions, on the other hand, distributes these values in a probably more natural way—at any rate, they provide greater flexibility. Figure A.3 shows a template for this class of functions.

The parameter θ allows interactions at all orders, while F is a probability distribution. Yet more flexibility is allowed in the choice of $+$ and $-$ signs: it is shown in [228] that these may have an important effect on the function's Hamming landscape. The only member of this class to be studied so far [228] is

$$F(\theta, k) = U(0, \theta^k)$$

with θ a scalar and $U(a, b)$ the uniform distribution over $a \le x \le b$. This would model a smooth decay (if $\theta < 1$), or a smooth increase (if $\theta > 1$) in the maximum (and mean) size of the interaction effects. (A smooth decay is usually assumed to be more likely in 'real' problems.) The signs in this case were chosen using a Bernoulli distribution with $p = 0.5$. This particular function also enables an easy calculation of the expected value of epistasis variance η.

Bibliography

[1] E.Aarts and J.K.Lenstra (Eds.) (1997) *Local Search in Combinatorial Optimization*, John Wiley & Sons, Chichester.

[2] A.Agapie (1998) Modeling genetic algorithms: from Markov chains to dependence with complete connections. In A.E.Eiben, T.Bäck, M.Schoenauer, H.P.Schwefel (Eds.) *Parallel Problem-Solving from Nature—PPSN V*, Springer-Verlag, Berlin, 3-12.

[3] A.Agapie (2001) Theoretical analysis of mutation-adaptive evolutionary algorithms. *Evolutionary Computation* **9**, 127-146.

[4] R.K.Ahuja and J.B.Orlin (1997) Developing fitter GAs. *INFORMS J on Computing*, **9**, 251-253.

[5] L.Altenberg (1995) The schema theorem and Price's theorem. In D.Whitley and M.Vose (Eds.) (1995) *Foundations of Genetic Algorithms 3*, Morgan Kaufmann, San Mateo, CA, 23-49.

[6] L.Altenberg (1997) Fitness distance correlation analysis: an instructive counter-example. In Th.Bäck (Ed.) (1997) *Proceedings of 7th International Conference on Genetic Algorithms*, Morgan Kaufmann, San Francisco, CA, 57-64.

[7] L.Altenberg (1997) NK fitness landscapes. In Th.Bäck, D.B.Fogel and Z.Michalewicz (Eds.) (1997) *Handbook of Evolutionary Computation*, Oxford University Press, Oxford, UK.

[8] E.Angel and V.Zissimopolous (2000) On the classification of NP-complete problems in terms of their correlation coefficients. *Discrete Applied Mathematics*, **99**, 261-277.

[9] J.Antonisse (1989) A new interpretation of schema notation that overturns the binary encoding constraint. In J.D.Schaffer (Ed.) (1989)

Proceedings of 3rd International Conference on Genetic Algorithms, Morgan Kaufmann, San Mateo, CA, 86-91.

[10] T.Bäck (1993), Optimal mutation rates in genetic search. In S.Forrest (Ed). *Proceedings of the Fifth International Conference on Genetic Algorithms*, Morgan Kaufmann, San Mateo, CA, 2-9.

[11] Th.Bäck (1995) Generalized convergence models for tournament and μ, λ selection. In L.J.Eshelman (Ed.) (1995) *Proceedings of 6th International Conference on Genetic Algorithms*, Morgan Kaufmann, San Francisco, CA, 2-8.

[12] Th.Bäck (1996) *Evolutionary Algorithms in Theory and Practice : Evolution Strategies, Evolutionary Programming, Genetic Algorithms*, Oxford University Press, Oxford.

[13] Th.Bäck, D.B.Fogel and Z.Michalewicz (Eds.) (1997) *Handbook of Evolutionary Computation*, Oxford University Press, Oxford, UK.

[14] Th.Bäck, A.E.Eiben and N.A.J.van der Vaart (2000) An empirical study on GAs "without parameters". M.Schoenauer, K.Deb, G.Rudolph, X.Yao, E.Lutton, J.J.Merelo and H-P.Schwefel (Eds.) (2000) *Parallel Problem-Solving from Nature, 6*, Springer-Verlag, Berlin, 315-324.

[15] J.E.Baker (1987) Reducing bias and inefficiency in the selection algorithm. In J.J.Grefenstette(Ed.) (1987) *Proceedings of the 2nd International Conference on Genetic Algorithms*, Lawrence Erlbaum Associates, Hillsdale, New Jersey, 14-21.

[16] J.M.Baldwin (1896) A new factor in evolution. *American Naturalist*, **30**, 441-451, 536-563.

[17] K.G.Beauchamp (1975) *Walsh Functions and Their Applications*. Academic Press, London.

[18] M.Behe (1996) *Darwin's Black Box: The Biochemical Challenge to Evolution*, Free Press, New York, NY.

[19] A.Bertoni and M.Dorigo (1993) Implicit parallelism in genetic algorithms. *Artificial Intelligence*, **61**, 307-314.

[20] A.D.Bethke (1981) *Genetic algorithms as function optimizers*. Doctoral dissertation, University of Michigan.

[21] H.G.Beyer (1996) Towards a theory of evolution strategies: Self-adaptation. *Evolutionary Computation*, **3**, 311-347.

[22] J.A.Bishop and L.M.Cook (1975) Moths, melanism and clean air. *Scientific American*, **232**, 90-99.

[23] T.Blickle and L.Thiele (1995) A mathematical analysis of tournament selection. In L.J.Eshelman (Ed.) *Proceedings of the Sixth International Conference on Genetic Algorithms*, Morgan Kaufmann, San Francisco, CA, 9-16.

[24] T.Blickle and L.Thiele (1996) A comparison of selection schemes used in evolutionary algorithms. *Evolutionary Computation*, **4**, 361-394.

[25] K.D.Boese, A.B.Kahng and S.Muddu (1994) A new adaptive multistart technique for combinatorial global optimizations. *Operations Research Letters*, **16**, 101-113.

[26] L.B.Booker (1987) Improving search in genetic algorithms. In L.Davis (Ed.) (1987) *Genetic Algorithms and Simulated Annealing*. Morgan Kaufmann, Los Altos, CA, 61-73.

[27] S.Bornholdt (1997) Probing genetic algorithm performance of fitness landscapes. In R.K.Belew and M.D.Vose (Eds.) (1997) *Foundations of Genetic Algorithms 4*, Morgan Kaufmann, San Francisco, CA, 141-154.

[28] G.E.P.Box and G.M.Jenkins (1970) *Time Series Analysis, Forecasting and Control*, Holden Day, San Francisco, CA.

[29] H.J.Bremermann (1962) *Optimization through Evolution and Recombination*, Spartan Books, Washington, DC.

[30] H.J.Bremermann, J.Rogson and S.Salaff (1966) Global properties of evolution processes. In H.H.Pattee (Ed.) (1966) *Natural Automata and Useful Simulations*, 3-42.

[31] C.L.Bridges and D.E.Goldberg (1987) An analysis of reproduction and crossover in a binary-coded genetic algorithm. In J.J.Grefenstette(Ed.) (1987) *Proceedings of the 2nd International Conference on Genetic Algorithms*. Lawrence Erlbaum Associates, Hillsdale, NJ, 9-13.

[32] A.Brown (1999) *The Darwin Wars*, Simon and Schuster, London.

[33] R.A.Caruana and J.D.Schaffer (1988) Representation and hidden bias: Gray vs. binary coding for genetic algorithms. *Proceedings of the 5th International Conference on Machine Learning.* Morgan Kaufmann, Los Altos, CA, 153-161.

[34] R.Cerf (1998) Asymptotic convergence of genetic algorithms. *Advances in Applied Probability* 30(2), 521-550.

[35] L.Chambers (Ed.) (1995) *Practical Handbook of Genetic Algorithms: Applications, Volume I*, CRC Press, Boca Raton, FL.

[36] L.Chambers (Ed.) (1995) *Practical Handbook of Genetic Algorithms: New Frontiers, Volume II*, CRC Press, Boca Raton, FL.

[37] S.Chen and S.Smith (1999) Putting the "genetics" back into genetic algorithms. In W.Banzhaf and C.R.Reeves (Eds.) (1999) *Foundations of Genetic Algorithms 5*, Morgan Kaufmann, San Francisco, CA, 103-116.

[38] P.C.Chu and J.E.Beasley (1998) Constraint handling in genetic algorithms: The set partitioning problem. *Journal of Heuristics*, **4**, 323-358.

[39] J.A.Coyne (1998) Not black and white. *Nature*, **396**, 35-36.

[40] G.A.Croes (1958) A method for solving travelling salesman problems. *Operations Research*, **6**, 791.

[41] J.C.Culberson (1995) Mutation-crossover isomorphisms and the construction of discriminating functions. *Evolutionary Computation*, **2**, 279-311.

[42] J.Culberson (1998) On the futility of blind search: An algorithmic view of "No Free Lunch". *Evolutionary Computation*, **6**, 109-127.

[43] J.M.Daida, S.P.Yalcin, P.M.Litvak, G.A.Eickhoff and J.A.Polito (1999) Of metaphors and Darwinism: deconstructing genetic programming's chimera. In P.J.Angeline (Ed.) (1999) *Proceedings of the 1999 Congress on Evolutionary Computation*, IEEE Press, Piscataway, NJ, 453-462.

[44] C.R.Darwin (1859) *On the Origin of Species by Means of Natural Selection or the Preservation of Favoured Races in the Struggle for Life*, John Murray, London.

[45] Y.Davidor (1990) Epistasis variance: suitability of a representation to genetic algorithms. *Complex Systems*, 4, 369-383.

[46] Y.Davidor (1991) Epistasis variance: a viewpoint on GA-hardness. In G.J.E.Rawlins (Ed.) (1991) *Foundations of Genetic Algorithms*, Morgan Kaufmann, San Mateo, CA, 23-35.

[47] L.Davis (Ed.) (1991) *Handbook of Genetic Algorithms*, Van Nostrand Reinhold, New York.

[48] L.D.Davis, D.Orvosh, A.Cox and Y.Qiu (1993) A genetic algorithm for survivable network design. In S.Forrest (Ed.) (1993) *Proceedings of 5th International Conference on Genetic Algorithms*, Morgan Kaufmann, San Mateo, CA, 408-415.

[49] T.E.Davis and J.C.Principe (1991) A simulated annealing like convergence theory for the simple genetic algorithm. In R.K. Belew and L.B. Booker (Eds.) *Proceedings of the fourth Int. Conf. on Genetic Algorithms*, Morgan Kauffman, San Mateo, CA, 174-181.

[50] T.E.Davis and J.C.Principe (1993) A Markov chain framework for the simple genetic algorithm. *Evolutionary Computation*, 1, 269-288.

[51] R.Dawkins (1982) *The Extended Phenotype*, Longman, London.

[52] R.Dawkins (1986) *The Blind Watchmaker*, Longman, London.

[53] R.Dawkins (1996) *Climbing Mount Improbable*, Viking, London.

[54] K.Deb and D.E.Goldberg (1993) Analyzing deception in trap functions. In L.D.Whitley (Ed.) (1993) *Foundations of Genetic Algorithms 2*, Morgan Kaufmann, San Mateo, CA, 93-108.

[55] K.A.De Jong (1975) *An analysis of the behavior of a class of genetic adaptive systems*, Doctoral dissertation, University of Michigan, Ann Arbor, Michigan. [Available as PDF files from http://cs.gmu.edu/~eclab/kdj_thesis.html.]

[56] K.A.De Jong (1992) Are genetic algorithms function optimizers? In R.Männer and B.Manderick (Eds.) (1992) *Parallel Problem-Solving from Nature, 2*, Elsevier Science Publishers, Amsterdam, 3-13.

[57] K.A.De Jong and W.M.Spears (1992) A formal analysis of the role of multi-point crossover in genetic algorithms. *Annals of Maths. and AI*, 5, 1-26.

[58] K.A.De Jong (1993) Genetic algorithms are NOT function optimizers. In D.Whitley (Ed.) (1993) *Foundations of Genetic Algorithms 2*, Morgan Kaufmann, San Mateo, CA, 5-18.

[59] K.A.De Jong, W.M.Spears and D.F.Gordon (1995) Using Markov chains to analyze GAFOs. In D.Whitley and M.Vose (Eds.) (1995) *Foundations of Genetic Algorithms 3*, Morgan Kaufmann, San Mateo, CA, 115-137.

[60] K.A.De Jong (2002) *Evolutionary Computation*, MIT Press, Cambridge, MA.

[61] D.J.Depew and B.H.Weber (1995) *Darwinism Evolving: Systems Dynamics and the Genealogy of Natural Selection*, MIT Press, Cambridge, MA.

[62] T.Dobzhansky (1962) *Mankind Evolving*, Yale University Press, New Haven, CT.

[63] S.Droste, T.Jansen and I.Wegener (1999) Perhaps not a free lunch but at least a free appetizer. In W.Banzhaf, J.Daida, A.E.Eiben, M.H.Garzon, V.Hanavar, M.Jakiela and R.E.Smith (Eds.) *Proceedings of the Genetic and Evolutionary Computation Conference*, Morgan Kaufmann, San Francisco, CA, 833-839.

[64] S.Droste, T.Jansen and I.Wegener (2001) Optimization with randomized search heuristics: The (A)NFL theorem, realistic scenarios and difficult functions. *J.Theoretical Computer Science*, (to appear).

[65] D-Z.Du and P.M.Pardalos (1998) *Handbook of Combinatorial Optimization: Volumes 1-3*, Kluwer Academic Publishers, Dordrecht, NL.

[66] M.Eigen (1971) Self-organization of matter and the evolution of biological macromolecules. *Die Naturwissenschaften*, **10**, 465-523.

[67] N.Eldredge and S.J.Gould (1973) Punctuated equilibria: An alternative to phyletic gradualism. In T.J.M.Schopf (Ed.) (1973) *Models in Paleobiology*, Freeman Cooper, San Francisco, CA, 82-115.

[68] T.M.English (1999) Some information theoretic results on evolutionary optimization. In P.J.Angeline (Ed.) (1999) *Proceedings of the 1999 Congress on Evolutionary Computation*, IEEE Press, Piscataway, NJ, 788-795.

[69] T.M.English (2000) Optimization is easy and learning is hard in the typical function. In *Proceedings of the 2000 Congress on Evolutionary Computation*, IEEE Press, Piscataway, NJ, 924-931.

[70] T.M.English (2000) Practical implications of new results in conservation of optimizer performance. In M.Schoenauer, K.Deb, G.Rudolph, X.Yao, E.Lutton, J.J.Merelo and H-P.Schwefel (Eds.) (2000) *Parallel Problem-Solving from Nature, 6*, Springer-Verlag, Berlin, 69-78,

[71] A.Eremeev and C.R.Reeves (2002) Non-parametric estimation of properties of combinatorial landscapes. To appear in J.Gottlieb and G.Raidl (Eds.) (2002) *Proceedings of EvoCOP-2002*, Springer-Verlag, Berlin.

[72] L.J.Eshelman, R.A.Caruana and J.D.Schaffer (1989) Biases in the crossover landscape. In J.D.Schaffer (Ed.) (1989) *Proceedings of 3rd International Conference on Genetic Algorithms*, Morgan Kaufmann, Los Altos, CA, 10-19.

[73] L.J.Eshelman (1991) The CHC adaptive search algorithm: How to have safe search when engaging in nontraditional genetic recombination. In G.J.E.Rawlins (Ed.) (1991) *Foundations of Genetic Algorithms*, Morgan Kaufmann, San Mateo, CA, 265-283.

[74] L.J.Eshelman and J.D.Schaffer (1993) Crossover's niche. In S.Forrest (Ed.) (1993) *Proceedings of 5th International Conference on Genetic Algorithms*, Morgan Kaufmann, San Mateo, CA, 9-14.

[75] A.Fairley (1991) *Comparison of methods of choosing the crossover point in the genetic crossover operation.* Technical Report, Dept. of Computer Science, University of Liverpool, UK.

[76] E.Falkenauer (1998) *Genetic Algorithms and Grouping Problems*, John Wiley & Sons, Chichester.

[77] T.C.Fogarty (1989) Varying the probability of mutation in the genetic algorithm. In J.D.Schaffer (Ed.) (1989) *Proceedings of 3rd International Conference on Genetic Algorithms*, Morgan Kaufmann, Los Altos, CA, 104-109.

[78] D.B.Fogel and J.W.Atmar (1990) Comparing genetic operators with Gaussian mutations in simulated evolutionary processes using linear systems. *Biological Cybernetics*, **63**, 111-114????

[79] D.B.Fogel (1998) *Evolutionary Computation: The Fossil Record*, IEEE Press, Piscataway, NJ.

[80] D.B.Fogel (1999) An overview of evolutionary programming. In L.D.Davis, K.A.De Jong, M.D.Vose and L.D.Whitley (Eds.) (1999) *Evolutionary Algorithms*: IMA Volumes in Mathematics and its Applications, Vol 111, Springer-Verlag, New York, 89-109.

[81] L.J.Fogel (1963) *Biotechnology: Concepts and Applications*, Prentice-Hall, Englewood Cliffs, NJ.

[82] L.J.Fogel, A.J.Owens and M.J.Walsh (1966) *Artificial Intelligence through Simulated Evolution*, John Wiley, New York.

[83] C.Fonlupt, D.Robilliard and P.Preux (1998) A bit-wise epistasis measure for binary search spaces. In A.E.Eiben, T.Bäck, M.Schoenauer, H.P.Schwefel (Eds.) *Parallel Problem-Solving from Nature—PPSN V*, Springer-Verlag, Berlin, 47-56.

[84] C.M.Fonseca and P.J.Fleming (1995) An overview of evolutionary algorithms in multiobjective optimization. *Evolutionary Computation*, **3**, 1-16.

[85] B.R.Fox and M.B.McMahon (1991) Genetic operators for sequencing problems. In G.J.E.Rawlins (Ed.) (1991) *Foundations of Genetic Algorithms*, Morgan Kaufmann, San Mateo, CA, 284-300.

[86] A.S.Fraser (1962) Simulation of genetic systems, *Journal of Theoretical Biology*, **2**, 329-346.

[87] P.Galinier and J.Hao (1999) Hybrid evolutionary algorithms for graph coloring. In *Journal of Combinatorial Optimization*, **3**, 379-397.

[88] M.R.Garey and D.S.Johnson (1979) *Computers and Intractability: A Guide to the Theory of NP-Completeness*, W.H.Freeman, San Francisco, CA.

[89] L.L.Gatlin (1972) *Information Theory and the Living System*, Columbia University Press, New York.

[90] H.Geiringer (1944) On the probability theory of linkage in Mendelian heredity. *Annals of Mathematical Statistics*, **15**, 25-57.

[91] F.Glover and M.Laguna (1993) Tabu Search. Chapter 3 in [213].

[92] F.Glover (1994) Genetic algorithms and scatter search: unsuspected potentials. *Statistics and Computing*, **4**, 131-140.

[93] D.E.Goldberg (1985) *Optimal initial population size for binary-coded genetic algorithms*. TCGA Report 85001, University of Alabama, Tuscaloosa.

[94] D.E.Goldberg and R.Lingle (1985) Alleles, loci and the traveling salesman problem. In J.J.Grefenstette(Ed.) (1985) *Proceedings of an International Conference on Genetic Algorithms and Their Applications*, Lawrence Erlbaum Associates, Hillsdale, New Jersey, 154-159.

[95] D.E.Goldberg and J.Richardson (1987) Genetic algorithms with sharing for multimodal function optimization. In J.J.Grefenstette(Ed.) (1987) *Proceedings of the 2nd International Conference on Genetic Algorithms*. Lawrence Erlbaum Associates, Hillsdale, NJ, 41-49.

[96] D.E.Goldberg (1989) *Genetic Algorithms in Search, Optimization, and Machine Learning*, Addison-Wesley, Reading, Massachusetts.

[97] D.E.Goldberg (1989) Sizing populations for serial and parallel genetic algorithms. In J.D.Schaffer (Ed.) (1989) *Proceedings of 3rd International Conference on Genetic Algorithms*, Morgan Kaufmann, San Mateo, CA, 70-79.

[98] D.E.Goldberg (1989) Genetic algorithms and Walsh functions: part I, a gentle introduction. *Complex Systems*, **3**, 129-152.

[99] D.E.Goldberg (1989) Genetic algorithms and Walsh functions: part II, deception and its analysis. *Complex Systems*, **3**, 153-171.

[100] D.E.Goldberg and K.Deb (1991) A comparative analysis of selection schemes used in genetic algorithms. In G.J.E.Rawlins (Ed.) (1991) *Foundations of Genetic Algorithms*, Morgan Kaufmann, San Mateo, CA, 69-93.

[101] D.E.Goldberg, K.Deb and J.H.Clark (1992) Genetic algorithms, noise, and the sizing of populations. *Complex Systems*, **6**, 333-362.

[102] B.Goodwin (1995) Neo-Darwinism has failed as an evolutionary theory. *Times Higher Education Supplement*, May 19 1995.

[103] B.Goodwin (1995) *How the Leopard Changed Its Spots*, Weidenfeld and Nicholson, London.

[104] S.J.Gould and R.Lewontin (1979) The spandrels of San Marco and the Panglossian paradigm: A critique of the adaptationist programme. *Proceedings of the Royal Society*, **B205**.

[105] R.L.Graham, D.E.Knuth and O.Patashnik (1994) *Concrete Mathematics* (2nd Edition), Addison-Wesley.

[106] J.J.Grefenstette (1986) Optimization of control parameters for genetic algorithms. *IEEE-SMC*, **SMC-16**, 122-128.

[107] J.J.Grefenstette (1993) Deception considered harmful. In [346], 75-91.

[108] G.R.Grimmett and D.R.Stirzaker (2001) *Probability and Random Processes*, 3rd edition, Oxford University Press, Oxford, UK.

[109] L.K.Grover (1992) Local search and the local structure of NP-complete problems. *Operations Research Letters*, **12**, 235-243.

[110] J.B.S.Haldane (1931) A mathematical theory of natural selection, Part VI: Metastable populations. *Proceedings of the Cambridge Philosophical Society*, **27**, 137-142.

[111] P.J.B.Hancock (1994) An empirical comparison of selection methods in evolutionary algorithms. In T.C.Fogarty (Ed.) (1994) *Evolutionary Computing: AISB Workshop, Leeds, UK, April 1994; Selected Papers*, Springer-Verlag, Berlin, 80-94.

[112] P.J.B.Hancock (1996) Selection methods for evolutionary algorithms. In L.Chambers (Ed.) (1995) *Practical Handbook of Genetic Algorithms: New Frontiers, Volume II*, CRC Press, Boca Raton, FL, 67-92.

[113] J.He and X.Yao (2001) Drift analysis and average time complexity of evolutionary algorithms. In *Artificial Intelligence*, **127**, 57-85.

[114] R.Heckendorn, S.Rana and D.Whitley (1997) Nonlinearity, hyperplane ranking and the simple genetic algorithm. In R.K.Belew and M.D.Vose (Eds.) (1997) *Foundations of Genetic Algorithms 4*, Morgan Kaufmann, San Francisco, CA, 181-202.

[115] R.B.Heckendorn and D.Whitley (1997) A Walsh analysis of NK-landscapes. In Th.Bäck (Ed.) (1997) *Proc. 7th International Conference on Genetic Algorithms*, Morgan Kaufmann, San Francisco, CA, 41-48.

[116] R.B.Heckendorn, D.Whitley and S.Rana (1997) Nonlinearity, hyperplane ranking and the simple genetic algorithm. In R.K.Belew and M.D.Vose (Eds.) (1997) *Foundations of Genetic Algorithms 4*, Morgan Kaufmann, San Francisco, CA, 181-201.

[117] R.B.Heckendorn and D.Whitley (1998) Predicting epistasis directly from mathematical models. Technical Report, Dept of Computer Science, Colorado State University, USA.

[118] R.Heckendorn, S.Rana and D.Whitley (1999) Test function generators as embedded landscapes. In W.Banzhaf and C.R.Reeves (Eds.) (1999) *Foundations of Genetic Algorithms 5*, Morgan Kaufmann, San Francisco, CA.

[119] K.Hinkelmann and O.Kempthorne (1994) *The Design and Analysis of Experiments*. John Wiley & Sons, New York.

[120] G.E.Hinton and S.J.Nowlan (1987) How learning can guide evolution. *Complex Systems*, 1, 495-502.

[121] C.Höhn and C.R.Reeves (1996) Graph partitioning using genetic algorithms. In G.R.Sechi (Ed.) (1996) *Proceedings of the 2nd Conference on Massively Parallel Computing Systems*, IEEE Computer Society, Los Alamitos, CA, 23-30.

[122] C.Höhn and C.R.Reeves (1996) The crossover landscape for the *Onemax* problem. In J.Alander (Ed.) (1996) *Proceedings of the 2nd Nordic Workshop on Genetic Algorithms and their Applications*, University of Vaasa Press, Vaasa, Finland, 27-43.

[123] C.Höhn and C.R.Reeves (1996) Are long path problems hard for genetic algorithms? In H-M.Voigt, W.Ebeling, I.Rechenberg and H-P.Schwefel (Eds.) (1996) *Parallel Problem-Solving from Nature—PPSN IV*, Springer-Verlag, Berlin, 134-143.

[124] J.H.Holland (1975) *Adaptation in Natural and Artificial Systems*, University of Michigan Press, Ann Arbor, Michigan.

[125] J.H.Holland (1986) Escaping brittleness: The possibilities of general-purpose learning algorithms applied to parallel rule-based systems. In R.S.Michalski, J.G.Carbonell and T.M.Mitchell (1986) *Machine Learning II*, Morgan Kaufmann, Los Altos, CA.

[126] J.H.Holland (1992) *Adaptation in Natural and Artificial Systems*, 2nd edition, MIT Press, Cambridge, MA.

[127] J.H.Holland (2000) Building blocks, cohort genetic algorithms, and hyperplane-defined functions. *Evolutionary Computation*, **8**, 373-391.

[128] J.Hooker (1995) Testing heuristics: We have it all wrong. *J.Heuristics*, **1**, 33-42.

[129] R.Hooykaas (1972) *Religion and the Rise of Modern Science*. Wm.B.Erdmans, Grand Rapids, Michigan.

[130] J.Horn, D.E. Goldberg, and K. Deb (1994) Long path problems. In Y.Davidor, H-P.Schwefel and R.Männer (Eds.) (1994) *Parallel Problem-Solving from Nature, 3*, Springer-Verlag, Berlin, 134-143.

[131] J.Horn and D.E.Goldberg (1999) Towards a control map for niching. In W.Banzhaf and C.R.Reeves (Eds.) (1999) *Foundations of Genetic Algorithms 5*, Morgan Kaufmann, San Francisco, CA, 287-310.

[132] C.Igel and M.Toussaint (2002) On classes of functions for which no free lunch results hold. Submitted to *IEEE Transactions on Evolutionary Computation*.

[133] T.Jansen and I.Wegener (2001) Real royal road functions—where crossover provably is essential. In L.Spector, E.D.Goodman, A.Wu, W.B.Langdon, H.-M.Voigt, M.Gen, S.Sen, M.Dorigo, S.Pezeshk, M.H.Garzon, E.Burke (Eds.) (2001) *Proceedings of the Genetic and Evolutionary Computation Conference (GECCO 2001)*, Morgan Kaufmann, San Francisco, CA, 1034-1041.

[134] D.S.Johnson (1990) Local optimization and the traveling salesman problem. In G.Goos and J.Hartmanis (Eds.) (1990) *Automata, Languages and Programming*, Lecture Notes in Computer Science **443**, Springer-Verlag, Berlin, 446-461.

[135] T.C.Jones (1995) *Evolutionary Algorithms, Fitness Landscapes and Search*, Doctoral dissertation, University of New Mexico, Albuquerque, NM.

[136] T.C.Jones (1995) Crossover, macromutation and population-based search. In L.J.Eshelman (Ed.) (1995) *Proceedings of the 6th International Conference on Genetic Algorithms*, Morgan Kaufmann, San Francisco, CA, 73-80.

[137] T.Jones and S.Forrest (1995) Fitness distance correlation as a measure of problem difficulty for genetic algorithms. In L.J.Eshelman (Ed.) *Proceedings of the 6th International Conference on Genetic Algorithms*, Morgan Kaufmann, San Mateo, CA, 184-192.

[138] B.Julstrom (1999) It's all the same to me: Revisiting rank-based probabilities and tournaments. In P.J.Angeline (Ed.) (1999) *Proceedings of the 1999 Congress on Evolutionary Computation*, IEEE Press, Piscataway, NJ, 1501-1505.

[139] L.Kallel and M.Schoenauer (1997) Alternative random initialization in genetic algorithms. In Th.Bäck (Ed.) (1997) *Proceedings of 7th International Conference on Genetic Algorithms*, Morgan Kaufmann, San Francisco, CA, 268-275.

[140] L.Kallel and J.Garnier (2001) *How to Detect All Maxima of a Function*. In L.Kallel, B.Naudts and A.Rogers (2001) *Theoretical Aspects of Evolutionary Computing*, Springer-Verlag, Berlin, 343-370.

[141] L.Kallel, B.Naudts and C.R.Reeves (2001) *Properties of Fitness Functions and Search Landscape*. In L.Kallel, B.Naudts and A.Rogers (2001) *Theoretical Aspects of Evolutionary Computing*, Springer-Verlag, Berlin, 175-206.

[142] L.Kallel, B.Naudts and A.Rogers (2001) *Theoretical Aspects of Evolutionary Computing*, Springer-Verlag, Berlin.

[143] A.Kapsalis, G.D.Smith and V.J.Rayward-Smith (1993) Solving the graphical Steiner tree problem using genetic algorithms. *J.Operational Research Society*, **44**, 397-406.

[144] S.Kauffman (1993) *The Origins of Order: Self-Organisation and Selection in Evolution*. Oxford University Press, Oxford, UK.

[145] S.Kauffman (1995) *At Home in the Universe*, Penguin, London.

[146] B.Kettlewell (1956) A résumé of investigations on the evolution of melanism in the Lepidoptera. *Proceedings of the Royal Society*, **B145**, 297-303.

[147] W.Khatib and P.J.Fleming (1998) The stud GA: A mini revolution? In A.E.Eiben, T.Bäck, M.Schoenauer, H.P.Schwefel (Eds.) *Parallel Problem-Solving from Nature, 5*, Springer-Verlag, Berlin, 683-691.

[148] E.L.Lawler, J.K.Lenstra, A.H.G.Rinnooy Kan and D.B.Shmoys (Eds.) (1990) *The Traveling Salesman Problem : A Guided Tour of Combinatorial Optimization*, John Wiley, Chichester.

[149] P.L'Ecuyer and C.Lemieux (2000) Variance reduction via lattice rules. *Management Science*, **46**, 1214-1235.

[150] J.Levenhagen, A.Bortfeldt and H.Gehring (2001) Path tracing in genetic algorithms applied to the multiconstrained knapsack problem. In E.J.W.Boers et al.(Eds.) (2001) *Applications of Evolutionary Computing*, Springer-Verlag, Berlin, 40-49.

[151] D.Levine (1997) GAs: A practitioner's view. *INFORMS J on Computing*, **9**, 256-257.

[152] S.Lin (1965) Computer solutions of the traveling salesman problem. *Bell Systems Tech. J.* ,**44**,2245-2269.

[153] S.L.Lohr (1999) *Sampling: Design and Analysis*, Duxbury Press, Pacific Grove, CA.

[154] M.Lundy and A.Mees (1986) Convergence of an annealing algorithm. *Mathematical Programming*, **34**, 111-124.

[155] W.G.Macready and D.H.Wolpert (1998) Bandit problems and the exploration/exploitation tradeoff. *IEEE Transactions on Evolutionary Computation*, **2**, 2-13.

[156] M.E.N.Majerus (1998) *Melanism: Evolution in Action* Oxford University Press, Oxford, UK.

[157] B.Manderick, M.de Weger and P.Spiessens (1991) The genetic algorithm and the structure of the fitness landscape. In R.K.Belew and L.B.Booker (Eds.) (1991) *Proceedings of 4th International Conference on Genetic Algorithms*, Morgan Kaufmann, San Mateo, CA, 143-150.

[158] O.Martin, S.W.Otto and E.W.Felten (1992) Large step Markov chains for the TSP incorporating local search heuristics. *Operations Research Letters*, **11**, 219-224.

[159] E.Mayr (1970) *Populations, Species, and Evolution*, Harvard University Press, Cambridge, Mass.

[160] E.Mayr (1982) *The Growth of Biological Thought: Diversity, Evolution, and Inheritance*, The Belknap Press of Harvard University Press, Cambridge, MA.

[161] M.D.McKay, W.J.Conover and R.J.Beckman (1979) A comparison of three methods for selecting values of input variables in the analysis of output from a computer code. *Technometrics*, **21**, 239-245.

[162] S.McKee and M.B.Reed (1987) An algorithm for the alignment of gas turbine components in aircraft. *IMA J Mathematics in Management*, **1**, 133-144.

[163] P.Merz and B.Freisleben (1998) Memetic algorithms and the fitness landscape of the graph bi-partitioning problem. In A.E.Eiben, Th.Bäck, M.Schoenauer, H-P.Schwefel (Eds.) (1998) *Parallel Problem-Solving from Nature—PPSN V*, Springer-Verlag, Berlin, 765-774.

[164] P.Merz and B.Freisleben (2000) Fitness landscape analysis and memetic algorithms for the quadratic assignment problem. *IEEE Transactions on Evolutionary Computation*, **4**, 337-352.

[165] M.M.Meysenburg and J.A.Foster (1997) The quality of pseudo-random number generators and simple genetic algorithm performance. In Th.Bäck (Ed.) (1997) *Proceedings of 7th International Conference on Genetic Algorithms*, Morgan Kaufmann, San Francisco, CA, 276-282.

[166] Z.Michalewicz (1996) *Genetic Algorithms + Data Structures = Evolution Programs* (3rd edition), Springer-Verlag, Berlin.

[167] Z.Michalewicz and M.Schoenauer (1996) Evolutionary algorithms for constrained parameter optimization problems. *Evolutionary Computation*, **4**, 1-32.

[168] M.Midgley (1985) *Evolution as a Religion : Strange Hopes and Stranger Fears*, Methuen, London.

[169] M.Mitchell, J.H.Holland and S.Forrest (1994) When will a genetic algorithm outperform hill climbing? In J.D.Cowan, G.Tesauro and J.Alspector (Eds.) (1994) *Advances in Neural Information Processing Systems 6*, Morgan Kaufmann, San Mateo, CA.

[170] M.Mitchell (1996) *An Introduction to Genetic Algorithms*, MIT Press, Cambridge, MA.

[171] J.B.Mitton (1997) *Selection in Natural Populations*, Oxford University Press, Oxford.

[172] D.C.Montgomery (1997) *Design and Analysis of Experiments*. Wiley, New York.

[173] H.Mühlenbein (1991) Evolution in time and space—the parallel genetic algorithm. In G.J.E.Rawlins (Ed.) (1991) *Foundations of Genetic Algorithms*, Morgan Kaufmann, San Mateo, CA, 316-337.

[174] H.Mühlenbein (1992) How genetic algorithms really work: mutation and hill-climbing. In R. Männer and B. Manderick (Eds.), *Parallel Problem Solving from Nature*, Elsevier Science Publishers, Amsterdam, 15-26.

[175] H.Mühlenbein and D.Schlierkamp-Voosen (1994) The science of breeding and its application to the breeder genetic algorithm. *Evolutionary Computation*, 1, 335-360.

[176] H.Mülenbein and T.Mahnig (2001) Evolutionary algorithms: from recombination to search distributions. In L.Kallel, B.Naudts and A.Rogers (Eds.) *Theoretical Aspects of Evolutionary Computing*, Springer-Verlag, Berlin, 135-173.

[177] Y.Nagata and S.Kobayashi (1997) Edge assembly crossover: A high-power genetic algorithm for the traveling salesman problem. In Th.Bäck (Ed.) (1997) *Proceedings of 7th International Conference on Genetic Algorithms*, Morgan Kaufmann, San Francisco, CA, 450-457.

[178] B.Naudts (1998) *Measuring GA-Hardness*, Doctoral thesis, University of Antwerp, RUCA, Belgium.

[179] B.Naudts and J.Naudts (1998) The effect of spin-flip symmetry on the performance of the simple GA. In A.E.Eiben, Th.Bäck, M.Schoenauer, H-P.Schwefel (Eds.) *Parallel Problem-Solving from Nature—PPSN V*, Springer-Verlag, Berlin, 67-76.

[180] B.Naudts and L.Kallel (2000) A comparison of predictive measures of problem difficulty in evolutionary algorithms. *IEEE Transactions on Evolutionary Computation*, 4, 1-15.

[181] A.Nijenhuis and H.S.Wilf (1978) *Combinatorial Algorithms for Computers and Calculators*. Academic Press, New York.

[182] A.E.Nix and M.D.Vose (1992) Modeling genetic algorithms with Markov chains. *Annals of Mathematics and Artificial Intelligence*, **5**, 79-88.

[183] C.E.Nugent, T.E.Vollman and J.E.Ruml (1968) An experimental comparison of techniques for the assignment of facilities to locations. *Operations Research*, **16**, 150-173.

[184] D.Orvosh and L.D.Davis (1993) Shall we repair? Genetic algorithms, combinatorial optimization and feasibility constraints. In S.Forrest (Ed.) (1993) *Proceedings of 5th International Conference on Genetic Algorithms*, Morgan Kaufmann, San Mateo, CA, 650.

[185] C.H.Papadimitriou and K.Steiglitz (1982) *Combinatorial Optimization: Algorithms and Complexity*. Prentice-Hall, Englewood Cliffs, NJ.

[186] C.Patterson (1999) *Evolution*, Comstock Press, Ithaca, NY.

[187] R.Poli (1999) Schema theorems without expectations. In W.Banzhaf, J.Daida, A.E.Eiben, M.H.Garzon, V.Hanavar, M.Jakiela and R.E.Smith (Eds.) *Proceedings of the Genetic and Evolutionary Computation Conference*, Morgan Kaufmann, San Francisco, CA.

[188] R.Poli (2001) Recursive conditional schema theorem, convergence and population sizing in genetic algorithms. In W.N.Martin and W.M.Spears (Eds.) (2001) *Foundations of Genetic Algorithms 6*, Morgan Kaufmann, San Francisco, CA, 143-163.

[189] R.Poli, J.E.Rowe and N.F.McPhee (2001) Markov chain models for GP and variable-length GAs with homologous crossover. In L.Spector, E.D.Goodman, A.Wu, W.B.Langdon, H.-M.Voigt, M.Gen, S.Sen, M.Dorigo, S.Pezeshk, M.H.Garzon, E.Burke (Eds.) (2001) *Proceedings of the Genetic and Evolutionary Computation Conference (GECCO 2001)*, Morgan Kaufmann, San Francisco, CA, 112-119.

[190] P.W.Poon and J.N.Carter (1995) Genetic algorithm crossover operators for ordering applications. *Computers & Operations Research*, **22**, 135-147.

[191] W.H.Press, S.A.Teukolsky, W.T.Vetterling and B.P.Flannery (1992) *Numerical Recipes in C : The Art of Scientific Computing*, Cambridge University Press, Cambridge, UK.

[192] G.R.Price (1970) Selection and covariance. *Nature*, **227**, 520-521.

[193] W.B.Provine (1971) *The Origins of Theoretical Population Genetics*, University of Chicago Press, Chicago.

[194] W.B.Provine (1986) *Sewall Wright and Evolutionary Biology*, University of Chicago Press, Chicago, IL.

[195] A.Prügel-Bennett and J.L.Shapiro (1997) The dynamics of a genetic algorithm for simple Ising systems, *Physica D*, **104**, 75-114.

[196] A.Prügel-Bennett and A.Rogers (2001) Modelling genetic algorithm dynamics. In L.Kallel, B.Naudts and A.Rogers (Eds.) *Theoretical Aspects of Evolutionary Computing*, Springer-Verlag, Berlin, 59-86.

[197] R.J.Quick, V.J.Rayward-Smith and G.D.Smith (1998) Fitness distance correlation and ridge functions. In A.E.Eiben, Th.Bäck, M.Schoenauer, H-P.Schwefel (Eds.) *Parallel Problem-Solving from Nature—PPSN V*, Springer-Verlag, Berlin, 77-86.

[198] N.J.Radcliffe (1991) Forma analysis and random respectful recombination. In R.K.Belew and L.B.Booker (Eds.) (1991) *Proceedings of 4th International Conference on Genetic Algorithms*, Morgan Kaufmann, San Mateo, CA, 222-229.

[199] N.J.Radcliffe (1991) Equivalence class analysis of genetic algorithms. *Complex Systems*, **5**, 183-205.

[200] N.J.Radcliffe (1992) Non-linear genetic representations. In R.Männer and B.Manderick (Eds.) (1992) *Parallel Problem-Solving from Nature, 2*, Elsevier Science Publishers, Amsterdam, 259-268.

[201] N.J.Radcliffe and F.A.W.George, 1993. A study in set recombination. In S.Forrest (Ed.) (1993) *Proceedings of 5th International Conference on Genetic Algorithms*, Morgan Kaufmann, San Mateo, CA, 23-30.

[202] N.J.Radcliffe (1994) The algebra of genetic algorithms. *Annals of Maths. and Artificial Intelligence*, **10**, 339-384.

[203] N.J.Radcliffe and P.D.Surry (1994) Formal memetic algorithms. In T.C.Fogarty (Ed.) (1994) *Evolutionary Computing: AISB Workshop, Leeds, UK, April 1994; Selected Papers*, Springer-Verlag, Berlin, 1-16.

[204] N.J.Radcliffe and P.D.Surry (1995) Formae and the variance of fitness. In D.Whitley and M.Vose (Eds.) (1995) *Foundations of Genetic Algorithms 3*, Morgan Kaufmann, San Mateo, CA, 51-72.

[205] N.J.Radcliffe (1997) Schema processing. In Th.Bäck, D.B.Fogel and Z.Michalewicz (Eds.) *Handbook of Evolutionary Computation*, Oxford University Press, Oxford, UK.

[206] N.J.Radcliffe and P.D.Surry (1997) Fundamental limitations on search algorithms: Evolutionary computing in perspective. In J.van Leeuwen (Ed.) (1997) *Lecture Notes in Computer Science 1000*, Springer-Verlag, Berlin, 275-291.

[207] S.Rana and D.Whitley (1997) Bit representations with a twist. In Th.Bäck (Ed.) (1997) *Proceedings of 7th International Conference on Genetic Algorithms*, Morgan Kaufmann, San Francisco, CA, 188-195.

[208] S.Rana (1999) *Examining the Role of Local Optima and Schema Processing in Genetic Search*. Doctoral thesis, Colorado State University, Fort Collins, CO.

[209] L.M.Rattray (1995) The dynamics of a genetic algorithm under stabilizing selection. *Complex Systems*, **9**, 213-234.

[210] L.M.Rattray and J.L.Shapiro (1996) The dynamics of a genetic algorithm for a simple learning problem. *Journal of Physics A*, **29**, 7451-7453.

[211] I.Rechenberg (1973) *Evolutionsstrategie: Optimierung technischer Systeme nach Prinzipen der biologischen Evolution*, Frommmann-Holzboog Verlag, Stuttgart. (2nd edition 1993).

[212] J.Rees and G.J.Koehler (1999) An investigation of GA performance results for different cardinality alphabets. In L.D.Davis, K.A.De Jong, M.D.Vose and L.D.Whitley (Eds.) (1999) *Evolutionary Algorithms*: IMA Volumes in Mathematics and its Applications, Vol 111, Springer-Verlag, New York, 191-206.

[213] C.R.Reeves (Ed.) (1993) *Modern Heuristic Techniques for Combinatorial Problems*, Blackwell Scientific Publications, Oxford, UK; re-issued by McGraw-Hill, London, UK (1995).

[214] C.R.Reeves (1994) Genetic algorithms and neighbourhood search. In T.C.Fogarty (Ed.) (1994) *Evolutionary Computing: AISB Workshop, Leeds, UK, April 1994; Selected Papers*, Springer-Verlag, Berlin, 115-130.

[215] C.R.Reeves (1994) Tabu search solves GA-hard problems in polynomial time. Presented at 15th International Symposium on Mathematical Programming, Ann Arbor, Michigan.

[216] C.R.Reeves (1995) A genetic algorithm for flowshop sequencing. *Computers & Operations Research*, **22**, 5-13.

[217] C.R.Reeves and C.C.Wright (1995) An experimental design perspective on genetic algorithms. In D.Whitley and M.Vose (Eds.) (1995) *Foundations of Genetic Algorithms 3*, Morgan Kaufmann, San Mateo, CA, 7-22.

[218] C.R.Reeves and C.C.Wright (1995) Epistasis in genetic algorithms: an experimental design perspective. In L.J.Eshelman (Ed.) (1995) *Proceedings of the 6th International Conference on Genetic Algorithms*, Morgan Kaufmann, San Mateo, CA, 217-224.

[219] C.R.Reeves (1996) Hybrid genetic algorithms for bin-packing and related problems. *Annals of Operations Research*, **63**, 371-396.

[220] C.R.Reeves (1996) Heuristic search methods: A review. In D.Johnson and F.O'Brien (1996) *Operational Research: Keynote Papers 1996*, Operational Research Society, Birmingham, UK, pages 122-149. ISBN 0903440164.

[221] C.R.Reeves (1997) Genetic algorithms for the Operations Researcher. *INFORMS J on Computing*, **9**, 231-250.

[222] C.R.Reeves and T.Yamada (1998) Genetic algorithms, path relinking and the flowshop sequencing problem. *Evolutionary Computation*, **6**, 45-60.

[223] C.R.Reeves (1999) Predictive measures for problem difficulty. In *Proceedings of 1999 Congress on Evolutionary Computation*, IEEE Press, 736-743.

[224] C.R.Reeves (1999) Landscapes, operators and heuristic search. *Annals of Operations Research*, **86**, 473-490.

[225] C.R.Reeves and C.C.Wright (1999) Genetic algorithms and the design of experiments. In L.D.Davis, K.A.De Jong, M.D.Vose and L.D.Whitley (Eds.) (1999) *Evolutionary Algorithms*: IMA Volumes in Mathematics and its Applications, Vol 111, Springer-Verlag, New York, 207-226.

[226] C.R.Reeves and T.Yamada (1999) *Goal-Oriented Path Tracing Methods*. In D.A.Corne, M.Dorigo and F.Glover (Eds.) (1999) *New Methods in Optimization*, McGraw-Hill, London.

[227] C.R.Reeves (2000) Fitness landscapes and evolutionary algorithms. In C.Fonlupt, J-K.Hao, E.Lutton, E.Ronald and M.Schoenauer (Eds.) (2000) *Artificial Evolution: 4th European Conference; Selected Papers*. Springer-Verlag, Berlin, 3-20.

[228] C.R.Reeves (2000) Experiments with tunable fitness landscapes. In M.Schoenauer, K.Deb, G.Rudolph, X.Yao, E.Lutton, J.J.Merelo and H-P.Schwefel (Eds.) (2000) *Parallel Problem-Solving from Nature—PPSN VI*, Springer-Verlag, Berlin, 139-148.

[229] C.R.Reeves (2001) Direct statistical estimation of GA landscape features. In W.N.Martin and W.M.Spears (Eds.) (2001) *Foundations of Genetic Algorithms 6*, Morgan Kaufmann, San Francisco, CA, 91-107.

[230] C.R.Reeves (2002) Estimating the number of optima in a landscape, Part I: Statistical principles. (In review).

[231] C.R.Reeves (2002) Estimating the number of optima in a landscape, Part II: Experimental investigations. (In review).

[232] C.R.Reeves (2002) The crossover landscape and the Hamming landscape for binary search spaces. To appear in *Foundations of Genetic Algorithms 7*, Morgan Kaufmann, San Francisco, CA.

[233] R.G.B.Reid (1985) *Evolutionary Theory: the Unfinished Synthesis*, Croom Helm, London.

[234] C.M.Reidys and P.F.Stadler (2002) Combinatorial landscapes. *SIAM Review*, **44**, 3-54.

[235] G.Reinelt (1994) *The Travelling Salesman: Computational Solutions for TSP Applications*, Springer-Verlag, Berlin.

[236] M.Ridley (1993) *Evolution*, Blackwell Scientific Press, Oxford, UK.

[237] S.M.Roberts and B.Flores (1966) An engineering approach to the travelling salesman problem. *Management Science*, **13**, 269-288.

[238] S.Rochet (1997) Epistasis in genetic algorithms revisited. *Information Science*.

[239] H.Rosé, W.Ebeling and T.Asselmeyer (1996) The density of states—a measure of the difficulty of optimization problems. In H-M Voigt, W.Ebeling, I.Rechenberg and H-P Schwefel (Eds.) (1996) *Parallel Problem-Solving from Nature—PPSN IV*, Springer-Verlag, Berlin.

[240] P.Ross (1997) srandom() anomaly. *Genetic Algorithms Digest*, http://www.www.aic.nrl.navy.mil/galist/ 11:23.

[241] P.Ross (1997) What are GAs good at? *INFORMS J on Computing*, **9**, 260-262.

[242] J.E.Rowe (1999) Population fixed-points for functions of unitation. In W.Banzhaf and C.R.Reeves (Eds.) (1999) *Foundations of Genetic Algorithms 5*, Morgan Kaufmann, San Francisco, CA, 69-84.

[243] J.E.Rowe (2001) A normed space of genetic operators with applications to scalability issues. *Evolutionary Computation*, **9**, 25-42.

[244] J.E.Rowe, M.D.Vose and A.H.Wright (2001) Group properties of crossover and mutation. *Evolutionary Computation*, **10**, 151-184.

[245] J.E.Rowe, M.D.Vose and A.H.Wright (2001) Structural search spaces and genetic operators. In preparation.

[246] G.Rudolph (1997) Reflections on bandit problems and selection methods in uncertain environments. In Th.Bäck (Ed.) (1997) *Proceedings of 7th International Conference on Genetic Algorithms*, Morgan Kaufmann, San Francisco, CA, 166-173.

[247] G.Rudolph (1997) *Convergence Properties of Evolutionary Algorithms*. Verlag Dr. Kovac, Hamburg, Germany.

[248] G.Rudolph (2000) Takeover times and probabilities of non-generational selection rules. In D.Whitley, D.E.Goldberg, E.Cantú-Paz, L.Spector, I.Parmee, and H.-G. Beyer (Eds.) (2000) *Proceedings of the Genetic and Evolutionary Computation Conference (GECCO 2000)*, Morgan Kaufmann, San Francisco, 903-910.

[249] G.Rudolph (2001) Takeover times of noisy non-generational selection rules that undo extinction. In V.Kůrková, N.C.Steele, R.Neruda and M.Kárný (Eds.) (2001) *Proceedings of the 5th International Conference on Artificial Neural Networks and Genetic Algorithms*, Springer-Verlag, Vienna, 268-271.

[250] E.Saliby (1990) Descriptive sampling: A better approach to Monte Carlo simulation. *Journal of Operational Research Society*, **41**, 1133-1142.

[251] R.Salomon (1999) The deterministic genetic algorithm: implementation details and some results. In P.J.Angeline (Ed.) (1999) *Proceedings of the 1999 Congress on Evolutionary Computation*, IEEE Press, Piscataway, NJ, 695-702.

[252] T.D.Sargent, C.D.Millar and D.M.Lambert (1998) The 'classical' explanation of industrial melanism: Assessing the evidence. In M.K.Hecht *et al.*(1998) *Evolutionary Biology*, **30**, 299-322, Plenum Press, New York.

[253] C.Schaffer (1994) A conservation law for generalization performance. In H.Willian and W.Cohen (Eds.) (1994) *Proceedings of 11th International Conference on Machine Learning*, Morgan Kaufmann, San Mateo, CA, 259-265.

[254] J.D.Schaffer, R.A.Caruana, L.J.Eshelman and R.Das (1989) A study of control parameters affecting online performance of genetic algorithms for function optimization. In J.D.Schaffer (Ed.) (1989) *Proceedings of 3rd International Conference on Genetic Algorithms*, Morgan Kaufmann, San Mateo, CA, 51-60.

[255] J.D.Schaffer and L.J.Eshelman (1996) Combinatorial optimization by genetic algorithms: The value of the genotype/phenotype distinction. In V.J.Rayward-Smith, I.H.Osman, C.R.Reeves and G.D.Smith (Eds.) (1996) *Modern Heuristic Search Methods*. John Wiley & Sons, Chichester, UK, 85-97.

[256] L.Schmitt, C.L.Nehaniv and R.H.Fujii (1998) Linear analysis of genetic algorithms. *Theoretical Computer Science*, **200**, 101-134.

[257] L.Schmitt (2001) Theory of genetic algorithms. *Theoretical Computer Science*, **259**, 1-61.

[258] N.N.Schraudolph and R.K.Belew (1992) Dynamic parameter encoding for genetic algorithms. *Machine Learning*, **9**, 9-21.

[259] C.Schumacher, M.D.Vose and L.D.Whitley (2001) The no free lunch theorem and description length. In L.Spector, E.Goodman, A.Wu, W.Langdon, H.-M.Voigt, M.Gen, S.Sen, M.Dorigo, S.Pezekh, M.Garzon and E.Burke (Eds.) (2001) *Proceedings of the Genetic and Evolutionary Computation Conference*, Morgan Kaufmann, San Francisco, CA, 565-570.

[260] H-P.Schwefel (1977) *Numerische Optimierung von Computer-modellen mittels der Evolutionsstrategie*, Birkhäuser Verlag, Basel. (English edition: *Numerical Optimization of Computer Models*, John Wiley & Sons, Chichester, 1981.)

[261] J.Shapiro (1997) A third way. *Boston Review*, **22(1)**, 32-33.

[262] J.Shapiro (2002) A 21st Century view of evolution, *J.Biol.Phys.*, to appear.

[263] J.Shapiro (2001) Statistical mechanics theory of genetic algorithms. In L.Kallel, B.Naudts and A.Rogers (Eds.) *Theoretical Aspects of Evolutionary Computing*, Springer-Verlag, Berlin, 87-108.

[264] J.Shapiro and A.Prügel-Bennett (1997) Genetic algorithm dynamics in a two-well potential. In R.K.Belew and M.D.Vose (Eds.) (1997) *Foundations of Genetic Algorithms 4*, Morgan Kaufmann, San Francisco, CA, 101-116.

[265] O.Sharpe (1998) On problem difficulty and further concerns with fitness-distance correlations. In *Symposium on Genetic Algorithms 1998*.

[266] S.S.Skiena (2000) *The Stony Brook Algorithm Repository*, http://www.cs.sunysb.edu/ algorith/index.html

[267] T.Soule and J.A.Foster (1997) Genetic algorithm hardness measures applied to the maximum clique problem. In Th.Bäck (Ed.) (1997) *Proc. 7th International Conference on Genetic Algorithms*, Morgan Kaufmann, San Francisco, CA, 81-88.

[268] W.M.Spears and K.A.De Jong (1997) Analyzing GAs using Markov models with semantically ordered and lumped states. In R.K.Belew

and M.D.Vose (Eds.) (1997) *Foundations of Genetic Algorithms 4*, Morgan Kaufmann, San Francisco, CA, 85-100.

[269] P.F.Stadler (1995) *Towards a Theory of Landscapes.* In R.Lopéz-Peña, R.Capovilla, R.García-Pelayo, H.Waelbroeck and F.Zertuche (Eds.) *Complex Systems and Binary Networks*, Springer-Verlag, Berlin, 77-163.

[270] P.F.Stadler and G.P.Wagner (1998) Algebraic theory of recombination spaces. *Evolutionary Computation*, **5**, 241-275.

[271] P.F.Stadler, R.Seitz and G.P.Wagner (2000) Population dependent Fourier decomposition of fitness landscapes over recombination spaces: Evolvability of Complex Characters *Bulletin of Mathematical Biology*, **62**, 399-428.

[272] C.R.Stephens and H.Waelbroeck (1997) Effective degrees of freedom in genetic algorithms and the block hypothesis. In Th.Bäck (Ed.) (1997) *Proceedings of 7th International Conference on Genetic Algorithms*, Morgan Kaufmann, San Francisco, CA, 34-40.

[273] C.R.Stephens and H.Waelbroeck (1999) Schemata evolution and building blocks. *Evolutionary Computation*, **7**, 109-124.

[274] C.R.Stephens, H.Waelbroeck and R.Aguirre (1999) Schemata as building blocks: does size matter? In W.Banzhaf and C.R.Reeves (Eds.) (1999) *Foundations of Genetic Algorithms 5*, Morgan Kaufmann, San Francisco, CA, 117-134.

[275] P.D.Surry and N.J.Radcliffe (1996) Inoculation to initialise evolutionary search. In T.C.Fogarty (Ed.) (1996) *Evolutionary Computing: AISB Workshop, Brighton, UK, April 1996; Selected Papers*, Springer-Verlag, Berlin, 269-285.

[276] J.Suzuki (1997) A further result on the Markov chain model. In R.K.Belew and M.D.Vose (Eds.) (1997) *Foundations of Genetic Algorithms 4*, Morgan Kaufmann, San Francisco, CA, 53-72.

[277] G.Syswerda (1989) Uniform crossover in genetic algorithms. In J.D.Schaffer (Ed.) (1989) *Proceedings of 3rd International Conference on Genetic Algorithms*, Morgan Kaufmann, Los Altos, CA, 2-9.

[278] D.Thierens (1997) Selection schemes, élitist recombination and selection intensity. In Th.Bäck (Ed.) (1997) *Proceedings of 7th International Conference on Genetic Algorithms*, Morgan Kaufmann, San Francisco, CA, 152-159.

[279] C.Thornton (1995) *Why GAs are Hard to Use.* Cognitive Science Research Paper 399, School of Cognitive and Computing Sciences, University of Sussex, UK.

[280] C.A.Tovey (1997) Local improvement on discrete structures. Chapter 3 in E.Aarts and J.K.Lenstra (Eds.) (1997) *Local Search in Combinatorial Optimization*, John Wiley & Sons, Chichester.

[281] P.Turney, D.Whitley and R.Anderson (Eds.) (1997) The Baldwin Effect. *Evolutionary Computation*, **4.3** (special issue).

[282] H.Van Hove and A.Verschoren (1995) On epistasis. *Computers and Artificial Intelligence*, **14**, 271-277.

[283] P.J.M.Van Laarhoven and E.H.L.Aarts (1988) *Simulated Annealing: Theory and Applications.* Kluwer, Dordrecht.

[284] E.Van Nimwegen, J.P.Crutchfield and M.Mitchell (1997) Finite populations induce metastability in evolutionary search. *Physics Letters A*, **229**, 144-150.

[285] M.D.Vose (1991) Generalizing the notion of a schema in genetic algorithms. *Artificial Intelligence*, **50**, 385-396.

[286] M.D.Vose and G.E.Liepins (1991) Schema disruption. In R.K.Belew and L.B.Booker (Eds.) (1991) *Proceedings of 4th International Conference on Genetic Algorithms*, Morgan Kaufmann, San Mateo, CA, 237-242.

[287] M.D.Vose (1993) Modeling simple genetic algorithms. In D.Whitley (Ed.) (1993) *Foundations of Genetic Algorithms 2*, Morgan Kaufmann, San Mateo, CA, 63-74.

[288] M.D.Vose (1994) A closer look at mutation in genetic algorithms. *Annals of Maths. and Artificial Intelligence*, **10**, 423-434.

[289] M.D.Vose and A.H.Wright (1994) Simple genetic algorithms with linear fitness *Evolutionary Computation*, **2**, 347-368.

[290] M.D.Vose (1995) Modeling simple genetic algorithms. *Evolutionary Computation*, **3**, 453-472.

[291] M.D.Vose and A.H.Wright (1995). Stability of vertex fixed points and applications. In D.Whitley and M.Vose (Eds.) (1995) *Foundations of Genetic Algorithms 3*, Morgan Kaufmann, San Mateo, CA, 103-114.

[292] M.D.Vose and A.H.Wright (1998) The simple genetic algorithm and the Walsh Transform: Part I, Theory. *Evolutionary Computation*, **6**, 253-273.

[293] M.D.Vose (1999) *The Simple Genetic Algorithm: Foundations and Theory*, MIT Press, Cambridge, MA.

[294] M.D.Vose (1999) What are genetic algorithms? A mathematical perspective. In L.D.Davis, K.A.De Jong, M.D.Vose and L.D.Whitley (Eds.) (1999) *Evolutionary Algorithms*: IMA Volumes in Mathematics and its Applications, Vol 111, Springer-Verlag, New York, 251-276.

[295] M.D.Vose (1999) Random heuristic search. *Theoretical Computer Science*, **229**, 103-142.

[296] M.D.Vose and L.D.Whitley (1999) A formal language for permutation recombination operators. In W.Banzhaf and C.R.Reeves (Eds.) (1999) *Foundations of Genetic Algorithms 5*, Morgan Kaufmann, San Francisco, CA, 135-146.

[297] M.D.Vose and A.H.Wright (2001) Form invariance and implicit parallelism. *Evolutionary Computation*, **9**, 355-370.

[298] R.A.Watson, G.S.Hornby and J.B.Pollack (1998) Modeling building-block dependency. In A.E.Eiben, T.Bäck, M.Schoenauer, H.P.Schwefel (Eds.) *Parallel Problem-Solving from Nature, 5*, Springer-Verlag, Berlin, 97-106.

[299] B.Weber (1999) Irreducible complexity and the problem of biochemical emergence. *Biology and Philosophy*, **14**, 593-605.

[300] E.D.Weinberger (1990) Correlated and uncorrelated landscapes and how to tell the difference. *Biological Cybernetics*, **63**, 325-336.

[301] D.Whitley (1989) The GENITOR algorithm and selection pressure: Why rank-based allocation of reproductive trials is best. In J.D.Schaffer (Ed.) (1989) *Proceedings of 3rd International Conference on Genetic Algorithms*, Morgan Kaufmann, Los Altos, CA, 116-121.

[302] D.Whitley (1991) Fundamental principles of deception in genetic search. In G.J.E.Rawlins (Ed.) (1991) *Foundations of Genetic Algorithms*, Morgan Kaufmann, San Mateo, CA, 221-241.

[303] D.Whitley, K.Mathias and P.Fitzhorn (1991) Delta coding: An iterative search strategy for genetic algorithms. In R.K.Belew and L.B.Booker (Eds.) (1991) *Proceedings of 4th International Conference on Genetic Algorithms*, Morgan Kaufmann, San Mateo, CA, 77-84.

[304] D.Whitley, T.Starkweather and D.Shaner (1991) The traveling salesman and sequence scheduling: Quality solutions using genetic edge recombination. In L.Davis (Ed.) (1991) *Handbook of Genetic Algorithms*, Van Nostrand Reinhold, New York, 350-372.

[305] D.Whitley(1992) Deception, dominance and implicit parallelism in genetic search. *Annals of Mathematics and Artificial Intelligence*, **5**, 49-78.

[306] D.Whitley (1993) An executable model of a simple genetic algorithm. In L.D.Whitley (Ed.) (1993) *Foundations of Genetic Algorithms 2*, Morgan Kaufmann, San Mateo, CA, 45-62.

[307] D.Whitley (1994) A genetic algorithm tutorial. *Statistics and Computing*, **4**, 65-85.

[308] D.Whitley, V.S.Gordon and K.Mathias (1994) Lamarckian evolution, the Baldwin effect and function optimization. In Y.Davidor, H-P.Schwefel and R.Männer (Eds.) (1994) *Parallel Problem-Solving from Nature, 3*, Springer-Verlag, Berlin, 6-15.

[309] D.Whitley and N-W.Yoo (1995) Modeling permutation encodings in the simple genetic algorithm. In D.Whitley and M.Vose (Eds.) (1995) *Foundations of Genetic Algorithms 3*, Morgan Kaufmann, San Mateo, CA, 163-184.

[310] D.Whitley and S.Rana (1997) Representation, search and genetic algorithms. In *Proceedings of 14th National Conference on Artificial Intelligence (AAAI-97)*, AAAI Press, Menlo Park, CA, 497-502.

[311] D.Whitley (1999) A free lunch proof for Gray versus binary encodings. In W.Banzhaf, J.Daida, A.E.Eiben, M.H.Garzon, V.Hanavar, M.Jakiela and R.E.Smith (Eds.) (1999) *Proceedings of the Genetic and Evolutionary Computation Conference*, Morgan Kaufmann, San Francisco, CA, 726-733.

[312] D.Whitley, L.Barbulescu and J-P.Watson (2001) Local search and high precision Gray codes: Convergence results and neighborhoods. In W.N.Martin and W.M.Spears (Eds.) (2001) *Foundations of Genetic Algorithms 6*, Morgan Kaufmann, San Francisco, CA, 295-311.

[313] S.W.Wilson (1991) GA-easy does not imply steepest-ascent optimizable. In R.K.Belew and L.B.Booker (Eds.) (1991) *Proceedings of 4th International Conference on Genetic Algorithms*, Morgan Kaufmann, San Mateo, CA, 85-89.

[314] D.H.Wolpert and W.G.Macready (1997) No free lunch theorems for optimization. *IEEE Transactions on Evolutionary Computation*, 1, 67-82.

[315] A.H.Wright and G.Bidwell (1997) A search for counterexamples to two conjectures on the simple genetic algorithm. In R.K.Belew and M.D.Vose (Eds.) (1997) *Foundations of Genetic Algorithms 4*, Morgan Kaufmann, San Francisco, CA, 73-84.

[316] A.H.Wright, R.K.Thompson and J.Zhang (2000) The computational complexity of $N - K$ fitness functions. *IEEE Transactions on Evolutionary Computation*, 4, 373-379.

[317] A.H.Wright and A.Agapie (2001) Cyclic and chaotic behaviour in genetic algorithms. In L.Spector, E.D.Goodman, A.Wu, W.B.Langdon, H.-M.Voigt, M.Gen, S.Sen, M.Dorigo, S.Pezeshk, M.H.Garzon, E.Burke (Eds.) (2001) *Proceedings of the Genetic and Evolutionary Computation Conference (GECCO 2001)*, Morgan Kaufmann, San Francisco, CA, 718-724.

[318] S.Wright (1932) The roles of mutation, inbreeding, crossbreeding and selection in evolution. In D.Jones (Ed.) (1932) *Proceedings of the 6th International Congress on Genetics*, 1, 356-366.

[319] F.Yates (1937) *Design and Analysis of Factorial Experiments*, Tech.Comm. No.35, Imperial Bureau of Soil Sciences, London.

[320] G.Zweig (1995) An effective tour construction and improvement procedure for the traveling salesman problem. *Operations Research*, **43**, 1049-1057.

Conference Proceedings

ICGA

[321] J.J.Grefenstette (Ed.) (1985) *Proceedings of an International Conference on Genetic Algorithms and their applications.* Lawrence Erlbaum Associates, Hillsdale, NJ.

[322] J.J.Grefenstette (Ed.) (1987) *Proceedings of the 2nd International Conference on Genetic Algorithms.* Lawrence Erlbaum Associates, Hillsdale, NJ.

[323] J.D.Schaffer (Ed.) (1989) *Proceedings of 3rd International Conference on Genetic Algorithms*, Morgan Kaufmann, San Mateo, CA.

[324] R.K.Belew and L.B.Booker (Eds.) (1991) *Proceedings of 4th International Conference on Genetic Algorithms*, Morgan Kaufmann, San Mateo, CA.

[325] S.Forrest (Ed.) (1993) *Proceedings of 5th International Conference on Genetic Algorithms*, Morgan Kaufmann, San Mateo, CA.

[326] L.J.Eshelman (Ed.) (1995) *Proceedings of 6th International Conference on Genetic Algorithms*, Morgan Kaufmann, San Mateo, CA.

[327] Th.Bäck (Ed.) (1997) *Proceedings of 7th International Conference on Genetic Algorithms*, Morgan Kaufmann, San Francisco, CA.

GECCO/CEC

[328] W.Banzhaf, J.Daida, A.E.Eiben, M.H.Garzon, V.Hanavar, M.Jakiela and R.E.Smith (Eds.) (1999) *Proceedings of the Genetic and Evolutionary Computation Conference (GECCO 1999)*, Morgan Kaufmann, San Francisco.

[329] D.Whitley, D.E.Goldberg, E.Cantú-Paz, L.Spector, I.Parmee, and H.-G. Beyer (Eds.) (2000) *Proceedings of the Genetic and Evolutionary Computation Conference (GECCO 2000)*, Morgan Kaufmann, San Francisco.

[330] L.Spector, E.D.Goodman, A.Wu, W.B.Langdon, H.-M.Voigt, M.Gen, S.Sen, M.Dorigo, S.Pezeshk, M.H.Garzon, E.Burke (Eds.) (2001) *Proceedings of the Genetic and Evolutionary Computation Conference (GECCO 2001)*, Morgan Kaufmann, San Francisco, CA.

[331] P.J.Angeline (Ed.) (1999) *Proceedings of the 1999 Congress on Evolutionary Computation*, IEEE Press, Piscataway, NJ.

[332] A.Zalzala (Ed.) (2000) *Proceedings of the 2000 Congress on Evolutionary Computation*, IEEE Press, Piscataway, NJ.

[333] J-H.Kim (Ed.) (2001) *Proceedings of the 2001 Congress on Evolutionary Computation*, IEEE Press, Piscataway, NJ.

PPSN

[334] H-P.Schwefel and R.Männer (Eds.) (1991) *Parallel Problem-Solving from Nature*, Springer-Verlag, Berlin.

[335] R.Männer and B.Manderick (Eds.) (1992) *Parallel Problem-Solving from Nature, 2*, Elsevier Science Publishers, Amsterdam.

[336] Y.Davidor, H-P.Schwefel and R.Männer (Eds.) (1994) *Parallel Problem-Solving from Nature, 3*, Springer-Verlag, Berlin.

[337] H-M.Voigt, W.Ebeling, I.Rechenberg and H-P.Schwefel (Eds.) (1996) *Parallel Problem-Solving from Nature, 4*, Springer-Verlag, Berlin.

[338] A.E.Eiben, T.Bäck, M.Schoenauer, H-P.Schwefel (Eds.) (1998) *Parallel Problem-Solving from Nature, 5*, Springer-Verlag, Berlin.

[339] M.Schoenauer, K.Deb, G.Rudolph, X.Yao, E.Lutton, J.J.Merelo and H-P.Schwefel (Eds.) (2000) *Parallel Problem-Solving from Nature, 6*, Springer-Verlag, Berlin.

ICANNGA

[340] R.F.Albrecht, C.R.Reeves and N.C.Steele (Eds.) (1993) *Proceedings of the International Conference on Artificial Neural Networks and Genetic Algorithms*, Springer-Verlag, Vienna.

[341] D.W.Pearson, N.C.Steele and R.F.Albrecht (Eds.) (1995) *Proceedings of the 2nd International Conference on Artificial Neural Networks and Genetic Algorithms*, Springer-Verlag, Vienna.

[342] G.D.Smith, N.C.Steele and R.F.Albrecht (Eds.) (1997) *Proceedings of the 3rd International Conference on Artificial Neural Networks and Genetic Algorithms*, Springer-Verlag, Vienna.

[343] A.Dobnikar, N.C.Steele, D.W.Pearson and R.F.Albrecht (Eds.) (1999) *Proceedings of the 4th International Conference on Artificial Neural Networks and Genetic Algorithms*, Springer-Verlag, Vienna.

[344] V.Kůrková, N.C.Steele, R.Neruda and M.Kárný (Eds.) (2001) *Proceedings of the 5th International Conference on Artificial Neural Networks and Genetic Algorithms*, Springer-Verlag, Vienna.

FOGA

[345] G.J.E.Rawlins (Ed.) (1991) *Foundations of Genetic Algorithms*, Morgan Kaufmann, San Mateo, CA.

[346] L.D.Whitley (Ed.) (1993) *Foundations of Genetic Algorithms 2*, Morgan Kaufmann, San Mateo, CA.

[347] D.Whitley and M.Vose (Eds.) (1995) *Foundations of Genetic Algorithms 3*, Morgan Kaufmann, San Mateo, CA.

[348] R.K.Belew and M.D.Vose (Eds.) (1997) *Foundations of Genetic Algorithms 4*, Morgan Kaufmann, San Francisco, CA.

[349] W.Banzhaf and C.R.Reeves (Eds.) (1999) *Foundations of Genetic Algorithms 5*, Morgan Kaufmann, San Francisco, CA.

[350] W.N.Martin and W.M.Spears (Eds.) (2001) *Foundations of Genetic Algorithms 6*, Morgan Kaufmann, San Francisco, CA.

Index

Author Index

Altenberg, Lee, 91, 262, 268

Bäck, Thomas, 60, 139
Behe, Michael, 12, 18
Bornholdt, Stefan, 197

Cerf, Raphael, 139
Culberson, Joseph, 109

Davidor, Yuval, 202, 228
Davis, Lawrence (Dave), 31, 45, 53
Davis, Thomas, 121, 138
Dawkins, Richard, 3, 11–14, 17
De Jong, Kenneth, 4, 5, 7, 39, 45, 46, 60, 131, 139
Deb, Kalyanmoy, 288

English, Tom, 109
Eremeev, Anton, 262

Falkenauer, Emanuel, 41, 60
Fogel, David, 2, 17
Fogel, Lawrence, 2, 17
Forrest, Stephanie, 228, 262
Freisleben, Bernd, 262

Gatlin, Lila, 13
Goldberg, David, 16, 17, 60, 91, 288
Goodwin, Brian, 13, 18
Gould, Stephen Jay, 11, 17

He, Jun, 284
Heckendorn, Robert, 228, 262
Höhn, Christian, 262
Holland, John, 1–4, 7, 10, 11, 16, 17, 60, 91

Jansen, Thomas, 197, 284
Jones, Terry, 228, 231, 262

Kallel, Leila, 221, 228, 262
Kauffman, Stuart, 13, 18, 262, 292

Macready, Bill, 109
Manderick, Bernard, 228
Mayr, Ernst, 4, 12, 15, 18
Merz, Peter, 262
Michalewicz, Zbigniew, 52, 58, 60
Midgley, Mary, 18
Mitchell, Melanie, 60, 91, 288
Mühlenbein, Heinz, 36, 55, 139

Naudts, Bart, 197, 221, 228, 262

Poli, Riccardo, 91, 170
Principe, Jose, 121, 138
Provine, William, 18
Prügel-Bennett, Adam, 197

Radcliffe, Nicholas, 29, 41, 53, 91, 109, 170
Rana, Soraya, 228, 262
Rattray, Magnus, 197
Rechenberg, Ingo, 2, 17
Reeves, Colin, 17, 228, 262

Rogers, Alex, 197, 262
Rowe, Jonathan, 170
Rudolph, Günter, 124, 138, 285

Schmitt, Lothar, 139
Schoenauer, Marc, 52
Schwefel, Hans-Paul, 2, 17
Shapiro, James, 12, 13
Shapiro, Jonathan, 197
Spears, William, 39, 131, 139
Stadler, Peter, 261
Stephens, Christopher, 91, 170, 268
Surry, Patrick, 29, 53, 109

Vose, Michael, 91, 138, 144, 169, 268

Wagner, Günter, 262
Watson, Richard, 290
Weber, Bruce, 13, 18
Wegener, Ingo, 197, 284
Weinberger, Edward, 262
Whitley, Darrell, 40, 46, 57, 60, 91, 109, 170, 228, 262
Wolpert, David, 109
Wright, Alden, 170, 262
Wright, Christine, 228
Wright, Sewall, 3, 14, 17

Yamada, Takeshi, 262
Yao, Xin, 284

Subject Index

allele, 20, 22, 29, 38, 42, 50, 56
alphabet, 19, 21, 27, 29, 65, 77
    minimal, 72, 91

Baldwin effect, 54
basin of attraction, 233
binary mask, 167
breeder GA, 55
building block, 71, 72, 266, 270, 277
    hypothesis, 57, 71, 78, 83, 265

Chapman-Kolmogorov equation, 112
characteristic function, 176
CHC algorithm, 57
chromosome, 65
commonality, 81
computational complexity, 284
constraints, 51
    penalty, 52
    repair, 52
convergence, 122–130
    almost surely, 124
convex hull, 146
coset, 166
crossover, 38
    HUX, 57
    bias, 38
    mask, 39, 40
    non-linear, 39
    one-point, 49
    operator, 158
crowding, 46
cumulants, 179
cumulants, 177
    finite population corrections, 190
    proportional selection, 181

selection and mutation, 189
selection, mutation and crossover, 195

Darwin, Charles, 2, 10, 17
deception, 78, 86, 88–90, 266, 267
deletion, 46
delta coding, 24
diagonalization, 154, 169
distance, 47
    Hamming, 47, 218, 222
distribution
    binomial, 182
    cumulants, 177
    Gaussian, 177, 183
    moments, 176
    multinomial, 119, 144
diversity, 30, 37, 44, 46, 48, 57
dynamic parameter encoding, 24
dynamics
    selection, 147
    selection + mutation, 156
    selection + mutation + crossover, 161

eigenvalue, 114, 151, 242
eigenvector, 114, 151, 242
élitism, 45, 57, 123
    convergence, 124
    threshold, 130
encoding, 20, 43
    binary, 21, 24
    discrete, 22
    permutation, 23
entropy, 198
epistasis, 202–217
    correlation, 215
    variance, 209
epistasis variance, 244
evolution strategy, 2, 25, 45, 57

experimental design, 204, 225
'exponential' growth, 75, 79

first hitting time, 284
Fisher, Sir Ronald, 17
fitness, 31
    in genetics, 14
    landscape, 14
    ratio, 68
    sharing, 46
fitness function, 14, 20, 33, 52
    NK landscape, 292
    ℓ, θ, 292
    $Onemax$, 125, 178, 288
    'basin with a barrier', 197
    deceptive, 290
    HIFF, 290
    long path, 291
    Royal Road, 78, 255, 288
    trap, 276, 288
    unitation, 115, 178, 287
fixed point, 257
    $Onemax$, 195
    asymptotically stable, 279
    hyperbolic, 162
    selection, 146
    selection + mutation, 152
focused heuristic, 162
force of $\mathcal{G}$, 152
forma, 81
Fourier series, 174
Fourier transform, 168

Geiringer's Theorem, 194
generation, 34, 49, 50, 53, 54, 58, 73, 76, 80
    gap, 45
generational GA, 45, 70
generational operator $\mathcal{G}$, 144
genetic drift, 147

genetic programming, 2, 170
genetics, 36
genotype, 4, 20, 21, 25, 30, 47, 49, 55
Gray code, 98
group
    commutative, 169
    transitive, 164
group action, 163
grouping states, 131

hitchhiking, 83
hybrid GAs, 52, 54
hyperplane, 65, 71, 74, 77, 83

implicit parallelism, 66, 76, 78, 265
incest prevention, 57
incremental GA, *see* steady-state GA
infinite population model, 144, 268
inner product, 88, 145, 168
intrinsic parallelism, *see* implicit parallelism
inversion, 43

knapsack problem, 21

Lamarckian evolution, 54
landscape, 163, 231–262
    amplitude spectrum, 243
    graph, 163, 240–245
limiting distribution, 121
local optimum, 89, 231, 233–241, 247, 250, 256–259
    formal definition, 239
local search, *see* neighbourhood search
locus, 3, 27, 43, 44, 50, 66

magnetization, 198
Markov chain, 111
    absorbing state, 116
    expected time to absorption, 117
    inhomogeneous, 127
Markov process, 111
mating, *see* crossover
matrix
    primitive, 113
    reducible, 116
    stochastic, 112
    transition, 112
metastable state, 154, 159, 172
mixing
    matrix, 84, 164, 169
    operator, 160, 164, 166, 169
multi-point crossover, 38
mutation, 30, 44–45, 49, 56, 57, 183–190
    adaptive, 31, 45, 129
    annealing, 127
    independent, 161
    matrix, 151
    rate, 31, 46, 50, 56, 57

neighbourhood, 105, 107, 232, 235
    CX, 235, 250
    bit-flip, 233
neighbourhood search, 7, 30, 54, 89, 231, 232
    first improvement, 233
    steepest ascent, 233
neutral networks, 162, 170, 269
niching, 46
'no duplicates', 32, 48
No Free Lunch theorem, 95–109, 273
NP-hard problems, 6, 214, 227

objective, 5, 21–23, 33, 35, 52, 100, 104
    multiple, 53

INDEX 331

order statistics, 284

permutation, 22, 39, 163
    matrix, 164
Perron-Frobenius theorem, 114, 152
phenotype, 4, 20, 21, 25, 30, 46, 47, 49, 55
population, 19, 45, 57
    finite, 161, 190
    initial, 25, 27, 29
    inoculation, 29
    size, 25, 27, 56
        variable, 58
    uniform, 119, 146, 257
    vector, 142, 173
Price's theorem, 80
probability generating function, 176
punctuated equilibrium, 269

quadratic
    form, 158
    operator, 158, 161

random number generation, 51
random search, 102, 104, 107, 108
recombination, *see* crossover
reduced surrogate, 48
replacement strategy, 45, 46
representation, 19–25, 163
respect, 81, 167
Robbins' proportions, 194

Santa Fe Institute, 169, 261
scaling, 33
schema, 65–72, 166, 267
    length, 67
    order, 67
schema theorem, 68, 70, 71, 168, 170
selection, 31–38, 119–120, 145–151, 179–183, 250, 265

'headless chicken', 250
    artificial, 15, 36, 55
    deletion, 53
    fitness-proportional, 31, 68, 73, 119, 135, 223
    intensity, 36, 55
    natural, 2, 11–13, 15
    population élitist, 57
    pressure, 33
    proportional, 145
    rank, 148
    ranking, 33, 46
    roulette wheel, 31, 49
    stochastic universal (SUS), 31
    strong, 128
    tournament, 34
    truncation, 36, 46, 56
sharing function, 46
simplex, 142
simulated annealing, 9, 54, 96, 127, 232, 260
statistical mechanics, 174
steady-state GA, 45, 48, 58, 123
stochastic process, 111
string, 19
    binary, 21, 23, 24, 27
    length, 20, 25, 27
structural
    crossover, 167
    space, 167
subgroup, 166
supermartingale, 124

tabu search, 52, 96, 106–108, 232, 260
takeover time, 37
termination, 30
two-armed bandit, 71–72, 77

Walsh coefficients, 87, 88, 207, 208,

210, 215, 217, 222, 224, 244, 245, 283
Walsh function, 87, 207, 244
Walsh transform, 86, 87, 168, 170, 282, 283

Books are to be returned on or before the last date below.

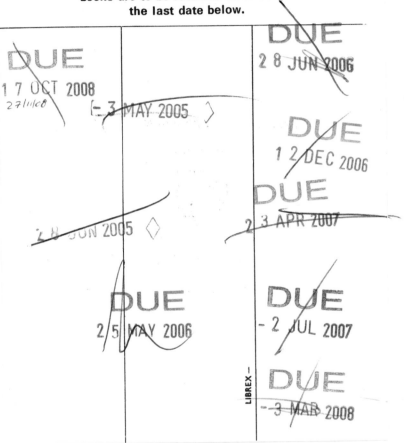